THE WORLD OF SCIENCE

INDEX

THE WORLD OF SCIENCE

INDEX

Facts On File Publications
New York, New York • Bicester, England

First published in the United States of America in 1986 by Facts on File, Inc., 460 Park Avenue South, New York, N.Y. 10016

First published in Great Britain in 1986 by Orbis Book Publishing Corporation Limited, London

Library of Congress Cataloging in Publication Data

Main entry under title:

World of Science

Includes index.
Summary: A twenty-five volume encyclopedia of scientific subjects, designed for eight-to twelve-year-olds. One volume is entirely devoted to projects.
1. Science—Dictionaries, Juvenile. 1. Science—Dictionaries
Q121.J86 1984 500 84–1654

ISBN: 0-8160-1077-3

Printed in Italy
10 9 8 7 6 5 4 3 2 1

Consultant editors
Eleanor Felder, Former Managing Editor, **New Book of Knowledge**
James Neujahr, Dean of the School of Education, City College of New York
Ethan Signer, Professor of Biology, Massachusetts Institute of Technology
J. Tuzo Wilson, Director General, Ontario Science Centre

Previous pages
Plants in desert areas have evolved different ways of surviving drought. The fleshy stems of cacti, for example, store water after rainstorms.

The computer featured as the front cover of this book was kindly supplied by Computech Systems Ltd., by courtesy of Apple Computer UK Ltd.
T-shirt courtesy of Motherart

Editor Penny Clarke
Designer Roger Kohn

► A sea anemone seen from above. The central mouth, surrounded by the rings of tentacles is clearly visible.

A

Aardvark **D** 23
Abacus **P** 13, **Y** 10–11
Aborigines **X** 8, 10, 11, 48, 49
Acanthodians **F** 19
Accelerator **P** 43
Accounting programs **Q** 38
Accumulator **Q** 13
Acetylene **K** 34
Acids **I** 19; **K** 36, 37, 43
Acoustic coupler **R** 45; **W** 13
Acropolis, Athens **J** 38
Actinosphaerium **S** 16
Acupuncture **I** 27
Adams, John Couch **N** 28
Adaptive radiation **G** 60
Adder **G** 39
Adding machine **P** 39; **U** 10
Addition **P** 12
Adenoids **H** 23
Ader, Clement **W** 48
Adrenalin **H** 29
Adrenals **H** 28
Advance Music System **R** 38
Adventure games **Q** 41; **R** 19
Aepyornis **F** 48
Aerials **U** 57
Aerobatics **R** 21
Aerodynamics
  aerofoil **U** 48
  drag **U** 48
  streamlining **U** 49
  testing **U** 48–49
Aetosaurs **F** 28
Afghanistan **X** 37
Africa
  people of **X** 10, 40–47
  savanna **A** 36
  tropical rainforests **A** 21
Agamemnon **W** 8
Age for earth
  ice ages **J** 28
  radiometric dating **J** 29
  relative dating **J** 29
  sea floor **J** 29
Age of Ancient Life **F** 13, 14
Age of Reptiles **C** 51
Ageing **H** 56–57
Agoraphobia **I** 38
Ahmes **P** 26, 27
Aiken, Howard **Q** 22
Ainu **X** 10
Air
  cleaning **M** 33
Air conditioning
  in cars **T** 24
  uses **T** 24
  working of system **T** 24
Aircraft navigation **W** 58–59
Airplane
  aileron **V** 7–9
  first flights **V** 7; **W** 48–52
  guiding **V** 7, 9
  jet engined **V** 9–11, 40–41
  make it and fly it **Y** 44, 45
  supersonic **V** 10
  turbofan **V** 11, 40
  turboprop **V** 40
  wings **V** 6, 8
Airports **W** 59
Airships **K** 9; **W** 46–47
Alaska **X** 13, 15
Alaska blackfish **C** 40
Albacore **C** 40
Albania **X** 31, 32
Albatross **B** 22
Albinism **I** 35
Alcock, John **W** 51, 56
Alder trees **G** 23
Aldrin, Edwin (Buzz) **V** 55
Aletsch glacier **J** 23
Alexander the Great **P** 29
Algae
  blue-green **A** 6; **F** 13, 17
  brown **A** 9
  features **A** 8
  fossilized **A** 6
  green **A** 8; **F** 20
  lichens **A** 56–57
  microscopic jungle **S** 18, 20
  red **A** 9
  reproduction **G** 14, 17
  suited to water **F** 20
  useful **A** 9
Algebra
  equations **P** 14–15
  higher **P** 41
  meaning **P** 14
  negative numbers **P** 15
  positive numbers **P** 15
  problems **P** 14
Algeria **X** 44
ALGOL **Q** 48
Alkalis **U** 17

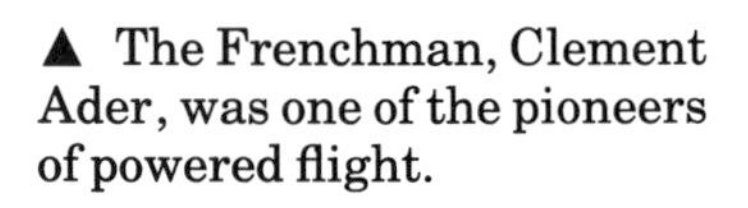

▲ The Frenchman, Clement Ader, was one of the pioneers of powered flight.

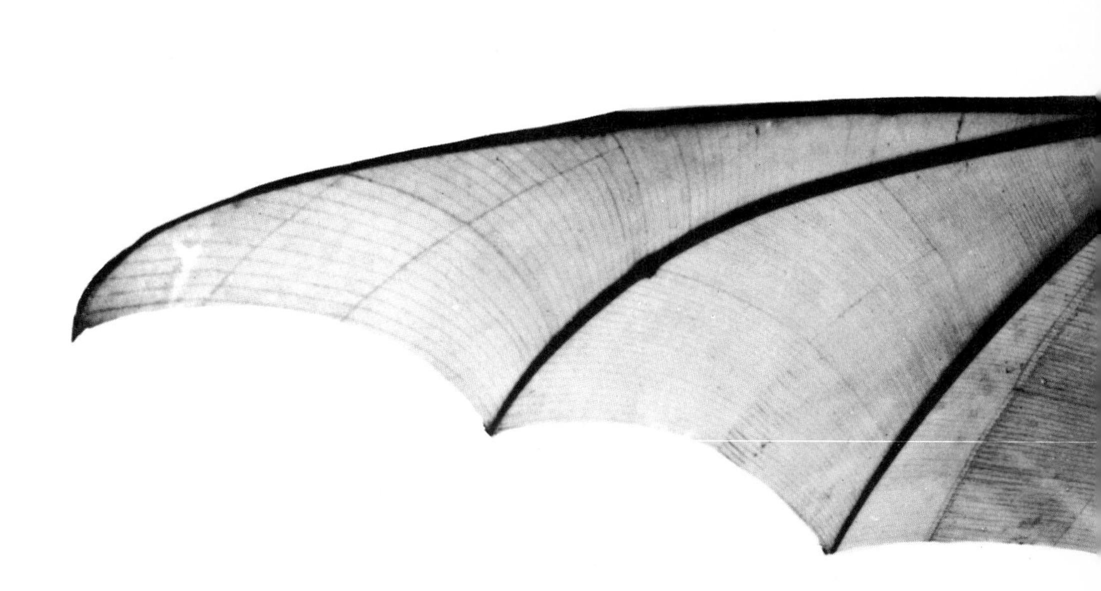

◀ *Avion III* was one of the airplanes designed by Ader. Built in 1897, it was powered by a steam engine, but it was uncontrollable and failed to take off.

Allergies **I** 36–37
Alligators **C** 50, 54
Allosaurus **F** 30, 31
Alloys **S** 34; **Y** 28, 32–33
Alpaca **D** 49
Alphabet **W** 7, 8
Alpine people **X** 27
Altering weather
by accident **M** 32–33
deliberately **M** 32
Alternation of generations **G** 16
Aluminum **J** 32, **Y** 32
Aluminum-magnesium alloy **Y** 32
Alveoli **S** 42
Amazon ants **E** 46
Amazon basin **A** 21, 22
Amazon jungle **S** 18, 20
Amazons **B** 39
Ambulances **I** 52
America
Central **X** 12, 13–14, 18
North **X** 10, 12, 13–14, 27, 55
South **X** 10, 20, 22–25
United States **X** 10, 13, 14, 15–16, 18, 22, 32, 33, 38
American Express **R** 47
American Indians **X** 10, 11, 13, 14, 15–16, 20, 22, 23, 25
American lobster **C** 30
Ammonites **F** 16–17
Amoeba **C** 8, 9; **G** 8, 11, 14, 49; **I** 33; **S** 16–17
Amphibians **C** 44–47; **F** 15, 19, 22–23; **G** 37–38
Amphipods **L** 34
Amplifiers **V** 48
Amstrad PCW 8256 **R** 24
Anabaena **S** 20
Analgesics **I** 27
Analogue signals **R** 36
Analytical engine **Q** 21
Anatomy
dissection **H** 7, 8
insect **E** 6, 10, 11
mammal **D** 10–11
meaning **H** 7, 8; **I** 14
modern aids **H** 8
origins of **H** 7
spider **E** 56
stag beetle **E** 48
Andorra **X** 30
Andromeda Nebula **N** 48
Anemometer **M** 30, 31
Anesthetic **I** 26–27, 59
Angel fish **C** 40
Angelica **A** 31
Angiosperms **A** 19; **F** 20
Anglaspis **F** 19
Anglerfish **C** 7
Angles **P** 18
Angola **X** 46
Animal
carriers of seeds **G** 23
game **Y** 14–15
plant-like **L** 30–31
reproduction **G** 26–47
Animation **R** 30–31
Ankylosaurus **F** 36
Annelid worm **S** 23
Anode attachment **Q** 15
Anorexia **I** 42
Ant-bear **D** 23
Antarctica **L** 15; **W** 54
Anteaters **D** 22, 23
Antelopes **D** 36, 58–59; **G** 53, 60; **X** 57
Antennae **E** 12
Anther **S** 8
Antibiotics **I** 24
Antibodies **H** 22, 27; **I** 21
Anticyclones
blocking **M** 26
droughts **M** 26
wind systems **M** 22
Antigen **I** 36
Antigua **X** 19
Antilles **X** 19
Antimatter **K** 30
Antiseptics **I** 22–23
Antiserum **S** 59
Ants
aphids and **E** 30–31
different kinds **E** 46
males **E** 46
nests **E** 46
pupae **E** 47
queens **E** 46
slave-making **E** 46
workers **E** 46, 47
Apartheid **X** 46
Apatosaurus **F** 30
Apes **D** 20–21
Aphids **E** 8–9, 30–31; **G** 13
Apianus, Petrus **N** 32
Apollo Mission **V** 53
Apollo Project **S** 34
Apollonius **P** 39
Appendicitis **I** 19
Appendix **H** 24–25
Applied mathematics **P** 35
Apricot F1 **R** 43

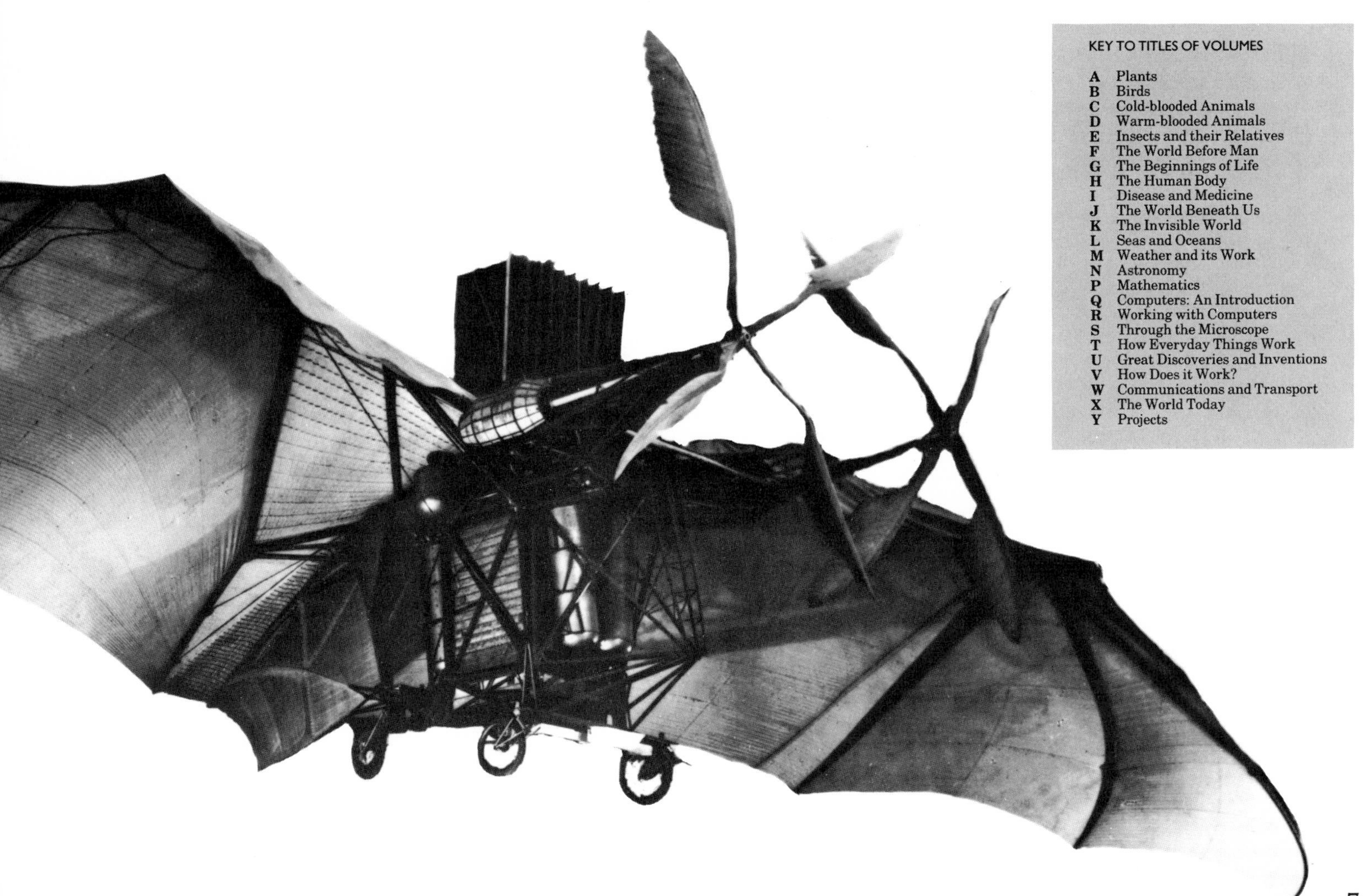

KEY TO TITLES OF VOLUMES

A Plants
B Birds
C Cold-blooded Animals
D Warm-blooded Animals
E Insects and their Relatives
F The World Before Man
G The Beginnings of Life
H The Human Body
I Disease and Medicine
J The World Beneath Us
K The Invisible World
L Seas and Oceans
M Weather and its Work
N Astronomy
P Mathematics
Q Computers: An Introduction
R Working with Computers
S Through the Microscope
T How Everyday Things Work
U Great Discoveries and Inventions
V How Does it Work?
W Communications and Transport
X The World Today
Y Projects

▲ Argentina: students at a school in north-west Argentina pose for a class photograph.

◄ Arctic animals, such as the ermine (**inset**) and Arctic fox, moult their dark summer coats in the autumn. Their winter coats are white – providing good camouflage against the snow.

Aquarium **Y** 59
Arabs **X** 10, 11, 35, 40, 44
Aratus of Soli **N** 36
Arcade games **Q** 40
Arcades **R** 19
Arcella **S** 14
Archaeopteryx **B** 8; **F** 41
Archaic Whites **X** 10, 48
Archelon **F** 38
Archimedes **P** 29; **Y** 27
Archosaurs **F** 28
Arctic **X** 55
Arctic Ocean **L** 8, 14, 15, 24
Argentavis **F** 48
Argentina **X** 22, 25
Argonaut **C** 22
Aristarchus of Samos **N** 8, 9
Aristotle **C** 18; **H** 7; **P** 29
Arithmetic
  basic **P** 7, 12–13
  higher **P** 41
  use of calculators **P** 13
Arithmetic and logic unit (ALU) **Q** 11, 12
Arkwright, Richard **U** 27
Armadillos **D** 22, 23
Armstrong, Neil **V** 55
Army ants **E** 46, 60
Arrays **Q** 33, 34
Arrow-worms **C** 12
Arrows **U** 6
Arteries **H** 18, 19
Arthritis **H** 16
Arthropods **C** 27–37; **F** 14, 22; **L** 33–34
Artificial breeding **G** 53
Artificial intelligence **Q** 47
Arts and math **P** 37
Asexual reproduction **G** 6, 8, 11, 13, 24
Asia
  people of **X** 9, 10, 32, 34–39, 55
  steppes **A** 36
  tropical rainforests **A** 21
Asparagus **A** 41
Aspergillus **S** 26–27
Asteroids **N** 10, 30
Astrolabe **P** 25, 34
Astrology
  mathematics muddled with **P** 31
Astronauts **K** 19, 46; **V** 2–3, 52
Astronomy (volume **N**)
  mathematics and **P** 24–25
Atari **R** 21
Athlete's foot **A** 11
Atlantic Ocean **L** 7, 8, 14, 16
Atmosphere **M** 6
Atom bomb **P** 8–9
Atomic nucleus **U** 13
Atoms **K** 6, 16, 20, 28–29; **S** 61; **U** 12–13, 54; **V** 22–23; **Y** 28
Audio/video connector **Q** 17
Audubon, John James **B** 56
Aurochs **D** 56
Aurora **Y** 37
Australia **X** 48, 49
Australian parakeets **Y** 57
Australoid **X** 10
Australopithecus **X** 6, 7
Austria **X** 30
Autogyros **V** 13; **W** 56–57
Automatic behaviour **K** 53
Automobile *see* Cars
Autopilot **V** 60
Axolotl **C** 45
Axons **H** 32–33
Aztecs **X** 13, 18; **Y** 12

KEY TO TITLES OF VOLUMES

A Plants
B Birds
C Cold-blooded Animals
D Warm-blooded Animals
E Insects and their Relatives
F The World Before Man
G The Beginnings of Life
H The Human Body
I Disease and Medicine
J The World Beneath Us
K The Invisible World
L Seas and Oceans
M Weather and its Work
N Astronomy
P Mathematics
Q Computers: An Introduction
R Working with Computers
S Through the Microscope
T How Everyday Things Work
U Great Discoveries and Inventions
V How Does it Work?
W Communications and Transport
X The World Today
Y Projects

B

B-52 jet bomber **V** 10
Babbage, Charles **P** 39; **Q** 20–21, 22
Baboon spider **C** 37
Babylonians **N** 6, 8
Bacon, Roger **P** 31
Bacteria **G** 53, 57; **H** 20; **I** 16, 19–21, 23; **K** 11; **S** 15, 25, 32, 37, 48–49, 52–53
Bacteriology **I** 16
Bacterium *see* Bacteria
Badgers **G** 56
Baekeland, Leo **U** 30
Bahamas **X** 19
Bahrain **X** 37
Baird, James Logie **T** 59; **W** 10, 11
Bakelite **U** 30–31
Balance **H** 38–39; **Y** 41
Ballooning **Y** 44
Balloons **W** 46–47
Baltic Sea **L** 10
Banded snail **C** 20
Bangladesh **X** 38
Baobab **G** 20
Bar codes **R** 53
Barbets **B** 40
Bardeen, John **Q** 25; **W** 10
Bark beetles **E** 50
Barn owl **Y** 35
Barnacles **E** 52; **G** 27; **L** 34
Barograph **M** 30
Barometer **M** 30
Barytes **J** 13
BASIC **Q** 13, 17, 28–29, 30–34, 48; **R** 14–15
Basking shark **C** 38
Basques **X** 29, 30
Bathrooms **S** 25
Bathysphere **L** 55
Bats **A** 32; **B** 6; **C** 61; **D** 18–19, 30, 62; **F** 55; **G** 20
Battery
  car **T** 50–51
  construction **T** 50–51
  types **T** 51
  working **V** 25, 36
Bauxite **J** 41
Bayer, Johann **N** 38
BBS *see* Bulletin Board Systems

◀ Birds' beaks differ according to the food they eat. The pelican's large beak and pouch are used to scoop up fish.

Beach buggy **U** 37
Beaches
  dunes **L** 20
  growth **L** 19
  groynes **L** 19
  longshore drift **L** 19
  pebble **L** 20
  spit **L** 20, 22
  see also Shores
Beaded lizard **C** 54–55
Beaks **B** 6
Beans **A** 42
Bears **D** 30–31; **G** 56
Beaufort, Sir Francis **M** 30
Beaufort Scale **M** 31
Bed bugs **E** 20, 21
Bee-eaters **B** 45
Beebe, William **L** 55
Beer **S** 37
Bees
  association with gardens **S** 29
  bumblebees **E** 43
  honeybee **E** 19, 40–42
  pollination **G** 19
  solitary **E** 43
Beetles
  body design **C** 34
  chrysalis **G** 33
  flesh-eating **E** 50–51
  foods **E** 50–51
  homes **E** 50
  identifying **E** 48, 50
  kinds **E** 48
  larvae **E** 15
  pet **Y** 58
Behaviour
  automatic **K** 53, 54
  born **K** 54–55
Belemnite **F** 17
Belgium **X** 28, 40
Belize **X** 18
Bell, Alexander Graham **W** 8
Benin **X** 45
Bennett's Comet **N** 32
Benz, Karl **W** 21, 22
Berbers **X** 10, 40, 44
Bernoulli, Daniel **U** 48
Bernoulli, Jacques **V** 7
Bernoulli's principle **V** 6, 7
Berries **G** 23
Bhutan **X** 38
Bicycles
  brakes **T** 8
  development of **T** 6–7
  folding **T** 8
  gears **T** 7–8, 10
  high-wheelers **T** 7
  penny-farthings **T** 7
  pumps **T** 14, 15
  races **T** 8
  racing **T** 8
  safety **T** 7
  stunt **T** 8–9
  tyres **T** 8
  velocipedes **T** 6
  working of **V** 33–34
Bile **H** 25
Bilharzia **S** 47
Binary fission **G** 6, 8, 11, 14, 49
Binary system **P** 11; **Q** 19; **R** 10, 32
Bingham Canyon, Utah **J** 40
Binocular vision **Y** 34, 35
Binoculars **T** 34; **Y** 22
Biological control **E** 18–19
Biologist **I** 54
Biped browsers **F** 34–35
Biplane **V** 7, 8
Birds (volume **B**)
  association with gardens **S** 29
  cold-blooded **C** 58–59
  early **F** 41
  early water **F** 47–48
  ecology **Y** 60

KEY TO TITLES OF VOLUMES

A	Plants
B	Birds
C	Cold-blooded Animals
D	Warm-blooded Animals
E	Insects and their Relatives
F	The World Before Man
G	The Beginnings of Life
H	The Human Body
I	Disease and Medicine
J	The World Beneath Us
K	The Invisible World
L	Seas and Oceans
M	Weather and its Work
N	Astronomy
P	Mathematics
Q	Computers: An Introduction
R	Working with Computers
S	Through the Microscope
T	How Everyday Things Work
U	Great Discoveries and Inventions
V	How Does it Work?
W	Communications and Transport
X	The World Today
Y	Projects

  internal fertilization **G** 28
  of open sea **L** 50
  on shore **L** 50
  pet **Y** 57, 60
  reproduction **G** 41–43
  species **G** 55
Birds-of-paradise **B** 60
Birds-of-prey **B** 31, 32
Bison **D** 56
Bits **Q** 26; **R** 10
Bittern **B** 10, 25; **Y** 35
Bivalve molluscs **C** 18–19, 24; **F** 16; **L** 37–38

SOME LESS WELL-KNOWN INVENTIONS

Reference books are full of 'firsts' – the first steam engine, the first printing machine, the first satellite, the first car and so on – and on! You'll have found the dates of many of these important inventions in the different volumes in this series. In the table below – and in similar tables throughout this volume – you'll find some less well-known 'firsts'. At first glance they may not seem so important, but without pencils or biros to write down the calculations and spanners to tighten the bolts, humans might still be struggling to reach the Moon!

BEFORE 1000BC

30,000	Paint in use in Europe for religious purposes (painting cave walls) by this date
5,000	Bricks in use in Middle East by this date; allowing construction of large buildings
3,000	Rope in use in south-west Asia
3,000	Ink in use in Egypt; it may have been used in China by an even earlier date
3,000	Beer popular in Egypt and Middle East
3,000	Dyes in use in India and Middle East
3,000	Adhesives used by Egyptians in making furniture
2,780	Solar (365 day) calendar in use in Egypt
2,500	Cement used by Egyptians in building the Pyramids
2,500	Asphalt in use in India and Middle East for waterproofing boats
1,300	Safety pins used in southern Europe to fasten heavy clothing
1,100	Umbrella in use as protection against rain or sun in China

KEY TO TITLES OF VOLUMES

A Plants
B Birds
C Cold-blooded Animals
D Warm-blooded Animals
E Insects and their Relatives
F The World Before Man
G The Beginnings of Life
H The Human Body
I Disease and Medicine
J The World Beneath Us
K The Invisible World
L Seas and Oceans
M Weather and its Work
N Astronomy
P Mathematics
Q Computers: An Introduction
R Working with Computers
S Through the Microscope
T How Everyday Things Work
U Great Discoveries and Inventions
V How Does it Work?
W Communications and Transport
X The World Today
Y Projects

Black ants **E** 46
Black bean aphids **E** 30
Black blight aphids **E** 30
Black Death **E** 16; **I** 13
Black rat **D** 26
Blackbird **B** 54
Blackbuck **D** 59
Blackfly **E** 8, 30, 38
Blacks **X** 10, 11, 15, 16, 22, 40, 44, 46
Bladder wrack **G** 14
Bladderworts **A** 55
Blanchard, Jean-Pierre **W** 47
Blériot, Louis **W** 48, 49
Blind people, aids for **R** 50–51
Block graphics **R** 29, 30, 32
Blood
  cells **K** 11; **S** 38
  functions **H** 20–21
  plasma **H** 20
  platelets **H** 21
  quantities **H** 20
  red cells **H** 15, 20
  supply **G** 47
  vessels **H** 19
  white cells **H** 20–21
Bloody-nosed beetle **C** 34
Blowflies **E** 38; **S** 26
Bluebottle fly **E** 13, 15, 38; **S** 26
Boa **C** 57
Boats **W** 34–38
Body
  cells **S** 41, 58–59
  chemicals **K** 40–43
  energy **K** 38–39
  louse **E** 21
  through the microscope **S** 38–39
  waves **K** 45
Boeing 747 jumbo jet **V** 6, 11
Bogs **A** 12
Bohr, Niels **U** 54
Boiling **K** 14
Bolivia **X** 20, 22, 23
Bolts
  construction **T** 18
  expansion **T** 18
Bones
  fossil **F** 7
  human **H** 16–17
Bonfires **W** 8
Bongo **G** 53
Bony fish **C** 40–43
Booklice **E** 21
Books **U** 24; **W** 7
Boole, George **Q** 25
Booth, Henry **W** 27
Borneo **X** 39
Botswana **X** 46
Bottlebrush **A** 19
Bowerbirds **B** 59
Box turtle **C**52
Brachiosaurus **F** 32, 33
Bracken **Y** 48
Bracket fungus **A** 10
Brahe, Tycho **N** 9, 16, 17, 42
Braille link **R** 50–51
Brain
  and intelligence **K** 50–53
  association fibres **H** 34
  automatic behaviour **K** 53
  cells **H** 9, 12
  cerebrum **H** 34, 35
  coral **C** 12
  and digestion **H** 50
  function **H** 34–35
  hypothalamus **H** 35
  medulla oblongata **H** 35
  moving **H** 44–45
  nervous system **H** 32
  olfactory bulbs **H** 35
  surgery **I** 10
  thalamus **H** 35
  tissues under the microscope **S** 42
  waves **K** 44–45
Braking system **V** 34
Brambles **G** 24
Branched sponge **C** 11
Brand, Commander Vance **W** 61
Brandenberger, Jacques **U** 29
Branly, Edouard **W** 9
Brass **Y** 32
Bratain, Walter **Q** 25; **W** 10
Brazil **X** 22, 23, 25
Bread **S** 37
Breadfruit **A** 21
Breastbone **B** 9
Breasts **H** 28, 52, 53
Breathing **E** 10–11
Breathing rate **B** 9
Breeding season **B** 13, 22
Breezes
  land **M** 22
  sea **M** 22
Breguet, Louis **V** 12
Bridges **W** 16, 17, 33
Brimstone butterfly **C** 34
Brine shrimps **S** 63
Bristlecone pine **A** 16
Bristles **D** 8
Bristletails **E** 20; **S** 25
Bristleworm **G** 27
Britain **X** 19, 22, 28, 29, 38, 40, 49, 50
Brittlestars **L** 35, 36; **Y** 20
Bromeliads **A** 22
Bronchiole **S** 42
Bronze Age **J** 30, 31; **Y** 32
Broom seeds **G** 22
Broomrapes **A** 58
Brown, Arthur Whitten **W** 51, 56
Brown, Sam **W** 20
Brown Lady of Raynam Hall, Norfolk **K** 56
Brunei **X** 39
Brunel, Isambard Kingdom **W** 39
Brush turkeys **B** 34
Bryce Canyon, Utah **J** 18
Buddhists **X** 39
Budding **C** 13; **G** 8
Budgerigars **B** 11, 39; **Y** 57
Buffalo **D** 56–57
Buffers **Q** 13
Bulbs **G** 25
Bulgaria **X** 31
Bulletin Board Systems (BBS) **R** 45–46
Bullfrog **C** 47
Bulls **Y** 37
Bumblebees **E** 43
Buoyancy **V** 50–51
Burdock seeds **A** 35
Burkina Faso **X** 45
Burma **X** 39
Burmese cats **Y** 56
Burnet moth **E** 34, 35
Burr marigold **G** 23
Burundi **X** 46
Burying beetle **E** 49, 50–51
Bush cricket **E** 28, 29
Bushmen **X** 8, 10, 44, 47
Bushnell, Nolan **Q** 40; **R** 19
Business communications **R** 45
Butler, Edward **W** 21
Butterflies **C** 34; **E** 9, 14, 15, 34–36; **G** 11, 19, 33; **S** 8, 29
Buzzards **B** 31
Bytes **Q** 13, 14, 26; **R** 10

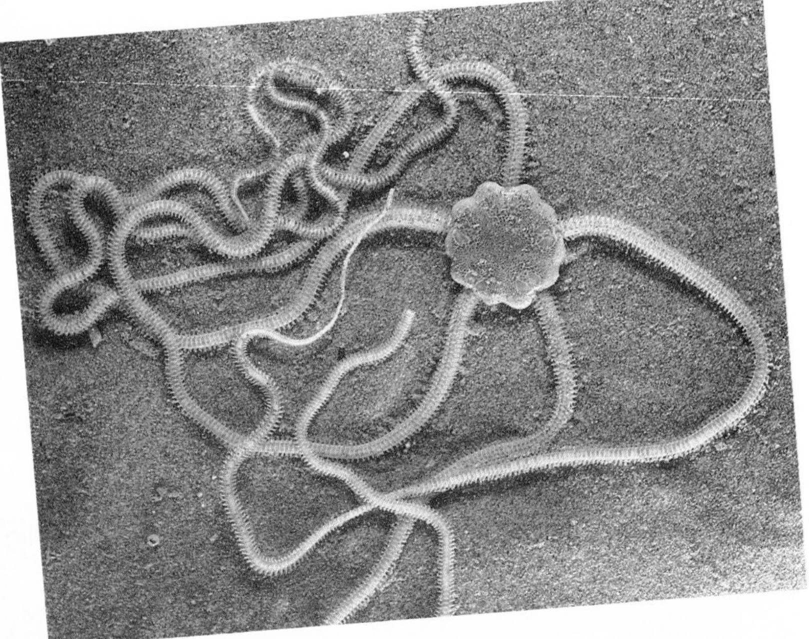

◄ Brittlestars often have very long arms. This species may have arms up to 18 cm (7 in) long.

► Brazil: the people living near this lagoon are of African descent – a reminder of the slave trade of past centuries.

C

Cabbage plants G 6
Cabbage root fly E 38
Cacomistle D 33
Cacti A 25, 52–53
Cadmium Y 32, 33
Caimans C 50
Calcite crystals Y 19
Calculating machines Q 19–21
Calculator
display P 13
invention P 38–39; U 10
microscopic parts S 37
using P 13
Calculus P 41
Calendar P 27; U 10
Calendar stone Y 12
Cambodia X 39
Cambrian period F 14
Camel spider C 37
Camels D 48–49
Cameras
automatic exposure control T 31
construction T 30–31
eyepieces T 32
film *see* Film
focusing T 30
instant T 33
lens T 33, 34
motion picture T 36–37
photo-electric cell T 31
processes involved T 32
self-focusing T 31
single-lens reflex T 30
television T 58
underwater L 54; T 31
viewfinders T 32
Cameroon X 45
Camptosaurus F 30
Canada X 13, 14, 15, 18
Canadian pond weed G 20
Canals L 54; W 44–45
Canaries B 11
Cancer H 58; I 7, 45–47, 60
Canning U 11
Cape Verde X 45
Čapek, Karel V 60
Capillaries H 19, 22
Capybara D 26, 27
Caracara B 31
Carbohydrates G 14; H 51; I 40, 41
Carbon Y 32
Carbon dioxide K 10; S 29
Carboniferous period F 14, 15, 20, 21; J 45
Carcinogens I 47
Carcinomas I 46
Caribou D 30, 50, 51
Carlsberg Caverns, New Mexico J 48
Carnivores F 51
Carnivorous plants A 54–55
Carnosaurs F 31, 33, 35
Carothers, Wallace U 31
Carpenter bee E 43
Carpet shark C 39
Cars
age of W 20–25
air conditioning T 24
computer uses R 40
engine V 35
petrol engine V 31
phones V 26
Cartilage H 16
Cartilaginous fish C 38; L 40–41
Cartoons R 21
Cartridge slot Q 17
Carts W 18, 19
Cassegrain, N N 52
Cassette
interface Q 10
port Q 17
recorder Q 6; R 9; T 47
tape recorder Q 54
unit Q 13
video T 61
Cassowary B 17
Castle, Keith I 60
Catbird B 53
Caterpillars C 34; E 15, 34; G 11, 33; S 9
Catfish Y 59
Cathode ray tube (CRT) Q 59; W 10
Catkins A 32
Cats
carrying babies B 15
classification G 56
in the wild D 36–37
pets Y 56
reproduction G 6
ringtailed D 33
Cattle egret B 25
Caucasoid X 10, 11, 15, 27, 35, 40, 48, 55
Cave-dwellers B 52
Caves J 48–49
Cayley, Sir George V 6; W 48, 56
Cedar of Lebanon A 15
Cellophane U 29
Cells
attacks on I 19
brain H 9, 12
cytoplasm H 12, 14
eye H 36
gland H 13
muscle H 31
nerve H 14; S 38, 39
nucleus H 12, 14
parts H 12
red H 16, 20
sex S 44
structures H 13
types H 12, 14–15
white H 20–21
Cellulose
celluloid U 29
fibre U 28
nitrocellulose U 29
pulp U 29
what is U 28–29
Celsius, Anders T 27
Cenozoic Era F 44–46
Census P 49
Centipedes
ancestors of F 22
distinguished from millipedes C 26–27; E 58, 59
female E 59
meaning C 26; E 58
mouth-parts E 58
ringed bodies C 34
size C 27
Central Africa X 10, 45
Central African Republic X 45
Central America X 12, 13–14, 18
Central Europe X 29–30
Central heating
hot air T 25
hot water T 25
Roman T 25
steam heating T 25
systems T 25
Central processing unit (CPU) P 56, 57; Q 10, 12
Centrifugal force U 40–41
Centrifuges
how they work U 40
what they do U 40
Cephalaspids F 19
Cephalopods F 16; L 38
Cepheid variables N 47
Ceramics
glaze U 20
heat-proof U 21

◀ Caterpillars of the lackey moth. The bright colours act as camouflage by breaking up the caterpillar's outline.

KEY TO TITLES OF VOLUMES

A Plants
B Birds
C Cold-blooded Animals
D Warm-blooded Animals
E Insects and their Relatives
F The World Before Man
G The Beginnings of Life
H The Human Body
I Disease and Medicine
J The World Beneath Us
K The Invisible World
L Seas and Oceans
M Weather and its Work
N Astronomy
P Mathematics
Q Computers: An Introduction
R Working with Computers
S Through the Microscope
T How Everyday Things Work
U Great Discoveries and Inventions
V How Does it Work?
W Communications and Transport
X The World Today
Y Projects

1000 BC TO AD 1000

700 False teeth of ivory, bone or human teeth made by the Etruscans in Italy
500 First known use of artificial leg
46 'Julian' calendar with leap years introduced in Rome

132 First seismograph invented in China
800 Paper money in use in China

1000 TO 1600

1280 Spectacles in use in Italy
1300 Italians make ice-cream (something rather similar had been made in more ancient times)
1400 Paint based on oil introduced
1509 'Pencils' in use: rods of lead-tin alloy
1550 Spanners for tightening nuts and bolts introduced
1565 First graphite pencil in wooden holder introduced
1582 Modern 'Gregorian' calendar introduced
1580s First modern newspaper, the *Mercurius Gallobelgicus*, published half-yearly in Germany

KEY TO TITLES OF VOLUMES

A Plants
B Birds
C Cold-blooded Animals
D Warm-blooded Animals
E Insects and their Relatives
F The World Before Man
G The Beginnings of Life
H The Human Body
I Disease and Medicine
J The World Beneath Us
K The Invisible World
L Seas and Oceans
M Weather and its Work
N Astronomy
P Mathematics
Q Computers: An Introduction
R Working with Computers
S Through the Microscope
T How Everyday Things Work
U Great Discoveries and Inventions
V How Does it Work?
W Communications and Transport
X The World Today
Y Projects

Ceratopsians **F** 37
Cereals
  cultivation **U** 19
  first grain growers **U** 18–19
  importance **A** 36–37
  machinery **U** 19
  production **U** 19
Cerebellum **S** 38, 39
Ceres **N** 10, 30
Ceylon *see* Sri Lanka
Chad **X** 44
Chadwick, James **U** 13
Chalk **Y** 19
Challenger mission **V** 58, 59
Challenger space craft **V** 20
Chameleons **C** 55; **Y** 58
Chaparral cock **B** 40
Chappe, Claude **W** 8
Character matrix **Q** 26
Character set **Q** 13
Chardonnet, Hilaire **U** 29, 30
Charles, J A C **Y** 44
Charles, Professor Jacques **W** 47
Charon **N** 29, 33
Cheetah **D** 36
Chemical makers **H** 28–29
Chemical tests **Y** 24–25
Chemical wastes **L** 61
Chemicals
  body **K** 40–43
  compounds **K** 36, 37
  discovery **I** 8
  elements **J** 8; **K** 36
  forces **K** 34–37
  reproduction **G** 6, 26
  substances **K** 36
Chemists **I** 54; **P** 7
Chemotherapy **I** 45
Chequered beetles **E** 50
Chesapeake Bay Bridge tunnel **W** 16
Chess-player **Q** 9
Chile **X** 20, 22, 23
Chimpanzees **D** 20
China **U** 21; **W** 14; **X** 31, 36, 56
Chinchillas **D** 9
Chinese **X** 15, 35, 49, 51; **Y** 29, 32
Chinese junks **W** 36
Chinese lantern **A** 35
Chinstrap penguins **B** 19
Chipmunks **D** 30
Chips (integrated circuits)
  clock timer **Q** 10
  EPROM **Q** 11, 13
  making **Q** 27
  meaning **Q** 11; **R** 10
  silicon **Q** 27; **W** 10
  sound **Q** 16
  support **Q** 13
  uses **U** 56
  video **Q** 17
Chiron **N** 10, 11, 29
Chitons **L** 38
Chlamydomonas **S** 20
Chlorine **K** 36
Chloroform **I** 27
Chlorophyll **A** 6, 10, 24; **K** 37; **L** 28; **S** 30
Chocolate **A** 43
Cholera **I** 60
Christmas tree **A** 14
Chromatography
  beginnings of **U** 16
  different kinds **U** 16
  meaning **U** 16
Chromium **Y** 32
Chromosomes **G** 49, 50; 58; **H** 54, 55; **I** 35; **K** 43; **S** 44–45
Chrysalis **E** 15, 34; **G** 33
Chuckawella Mountains **C** 58
Cicadas **E** 13, 28, 29
Cichlids **G** 36, 60
Cider making **Y** 46–47
Cierva, Juan de la **V** 13; **W** 56
Ciliate protozoan **Y** 54
Ciliates **C** 8, 9; **S** 16
Circulation system **H** 11, 18, 20
Cities, development of **U** 8; **X** 56, 57
Citrus fruits **A** 47
Civets **D** 35
Clams **C** 18, 19
Classifications **G** 56

▼ Chess, one of the most ancient of games, is believed to have originated in China. Skill and great concentration are required to become a good player, though there are now computer programs against which you can test your skill at the game.

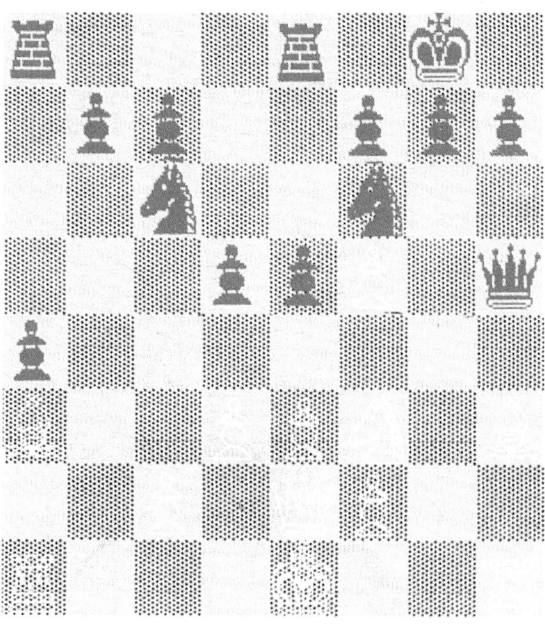

Classifying birds **B** 15
Cleaner wrasse **G** 13
Cleavers **G** 23
Click beetles **E** 50
Cliffs
  rocky pillars **L** 18
  sea's attack **L** 18–19
  stacks **L** 18
  types **L** 19
Climate
  climatic regions of world **M** 34–35
  cold forest **M** 48–49
  desert **M** 44–45
  ingredients **M** 35
  living things and **M** 35
  monsoon **M** 40–41
  mountain **M** 52–53
  polar **M** 50–51
  temperate **M** 46–47
  tropical rainy **M** 38–39
Clippers **W** 38
Clockmaking **U** 10
Clocks
  atomic **T** 53
  computers **Q** 18
  construction **T** 53
  early **Y** 13
Clones **G** 49
Closterium **S** 20
Cloth industry **Q** 19
Clothes **U** 6
Clouds
  altocumulus **M** 12
  cirrus **M** 12, 13
  condensation particles **M** 10
  cumulonimbus **M** 12–13
  feathery **M** 12
  formation **L** 6
  growth **M** 10
  heaped **M** 12
  high **M** 12
  ice **M** 10
  layered **M** 12
  low **M** 12
  medium **M** 12
  mountain **M** 53
  nimbostratus **M** 12
  stratocumulus **M** 12
  thundercloud **M** 11, 16, 17
  types **M** 12–13
Clytemnestra, Queen **K** 56; **W** 8
Co-evolution **G** 60
Coaches **W** 19
Coal **J** 45–46; **K** 24
Coatimundi **D** 33
Cobra **C** 56
Cockatoo **B** 39
Cockcroft, John **P** 43
Cockles **C** 18, 19
Cockroaches **E** 8; **S** 25
Cocoa **A** 43
Coconuts **A** 44–45; **G** 23
Cod **G** 34
Codes **Y** 16–17
Coelacanth **F** 19
Coelenterates **L** 30
Coelophysis **F** 31
Coelurosaurs **F** 31
Coffee **A** 43–44

Coherer **W** 9
Cold-blooded animal (volume **C**)
meaning **C** 6–7; **D** 6, 7
Cold-blooded birds **C** 58–59
Colds **S** 48
Collecting
fossils **Y** 20–21
minerals **Y** 18–19
Colonies **B** 23
Colorado, USA **Y** 44
Colorado beetle **E** 51
Colour
experiments with **Y** 36–37
monitor **Q** 14–15
vision **B** 10–11
Columbia **V** 18; **X** 22
Columbus, Christopher **U** 34; **W** 36
Combine harvester **U** 18, 19
Comets **N** 10, 11, 32
Commodore 64 **R** 31, 38
Communicating with computers **R** 40–51
Communications volume **W**
Communism **X** 27, 31–33, 36, 39
Compasses **L** 53; **P** 36, 37; **U** 42; **Y** 29, 32
Complete metamorphosis **C** 34; **E** 14, 15; **G** 33
Compounds, chemical **K** 36, 37
Compressed air **K** 19
Computers (volumes **Q** and **R**)
building **Y** 10–11
communications **W** 60
competing against **P** 57
dangerous **P** 57–58
earliest attempts **Q** 18
electronic circuits **T** 55
first **Q** 22–25
flow charts **P** 54
games **P** 58
gigo gremlin **P** 59–60
in medicine **I** 54, 61
in use **Q** 44–47
information for **P** 54–55
languages **P** 54
mainframe **P** 55, 59; **Q** 8, 9; **R** 18; **Y** 14
meaning **Q** 6, 7
micro **P** 56, **Y** 11
microscopic parts **S** 37
minis **P** 55–56
new uses **P** 60–61
parts **Q** 9
peripherals **Q** 54–59
print out **P** 55
program **P** 54
programmer **P** 54
projects **Y** 12–15
railway engineering **N** 13
system **P** 56
talking **P** 56–57
telephones **W** 32
uses **P** 53–54
working of **V** 27, 60
Concept keyboards **R** 51
Concorde **V** 8, 10, 11, 51
Condensation
humidity **M** 8
particles **M** 10
Condor **B** 29
Cone-bearing trees **A** 6–7, 14–15; **G** 17
Congo **X** 45
Coniferous trees *see* Cone-bearing trees
Conservation
animals **X** 57
of the Earth **X** 60
seas **L** 60–61
wild plants **A** 60–61
Consonants **W** 6
Constellations
Key **N** 37
observation of **N** 38
origins **N** 36–37
Continental plates **J** 10, 23
Control block **Q** 12
Controlled explosions **J** 35
Controls **Q** 14
Convection cells **J** 8, 10
Convergent evolution **G** 60
Cook, Captain James **C** 52
Cook Island **X** 51
Cooke, William **W** 8
Cooksonia **F** 20
Copepods **S** 22, 33; **Y** 54
Copernicus, Nicolaus **N** 9, 16, 17; **P** 35
Copper **J** 31, 32; **U** 8; **Y** 32
Coral
atolls **L** 35
brain **C** 12; **L** 2–3
close-up **S** 12–13
first appearance **G** 54
island lagoon **L** 25
polyps **L** 24, 25, 30
reefs **C** 13; **F** 17; **G** 8; **J** 24; **L** 24, 25, 61
Coriolis, Caspard de **L** 12
Cormorants **B** 24; **G** 59
Corms **G** 25
Corn marigolds **A** 7
Cornu, Paul **V** 12
Coryphodon **F** 46
Cosmic rays **J** 12; **K** 22
Costa Rica **X** 18
Cote d'Or region, France **Y** 46
Cotyledons **A** 19; **G** 21
Cotylosaurs **F** 24–25
Cougar **D** 36
Counting **Q** 26
Courtship **B** 21; **G** 27–28
Cousteau, Jacques **L** 55
Coutell, Captain **W** 47
Cowbirds **B** 13
Cowpox **I** 17, 18
Coyote **D** 29
CPU *see* Central processing unit
Crab louse **E** 21
Crabs **C** 24, 28, 31–32, 34; **E** 52, 53–54; **F** 18; **L** 34
Cranefly **E** 38, 39
Cranes **B** 35
Crayfish **C** 30; **E** 53
Creodonts **F** 51
Crested newt **C** 45
Cretaceous times **F** 28
Crickets **E** 13, 28, 29
Crocodiles **C** 49–50; **F** 29, 38; **G** 38, 39; **L** 50
Crocus **G** 25
Cross-pollination **G** 19
Crossbills **A** 15
Crown-of-thorns starfish **C** 25
Crows **B** 11, 61
CRT *see* Cathode ray tube
Cruise liners **W** 41
Crustaceans
bodies **C** 34
decapods **E** 53–54
food for fish **E** 52
homes **E** 52
land **E** 54
sub-group **C** 28
swimming **L** 33
Crystal set **W** 9
Crystals **J** 13; **K** 16, 17; **S** 34–36
Cuba **X** 19
Cubes **P** 16
Cuckoos **B** 13, 40
Cugnot, Nicolas **W** 20
Cuneiform **W** 7
Curlew **B** 36
Currency **P** 11
Currents
Canaries **L** 12
Coriolis effect **L** 12
deep-sea **L** 12
drifts **L** 12
gyres **L** 12
Labrador **L** 12
North Atlantic **L** 12
surface **L** 11, 12
underwater **L** 11
Curtiss, Glenn **W** 50
Cuttlefish **C** 18, 22; **G** 28
Cuvier, Georges **F** 9
Cycads **A** 15
Cyclamen **A** 29
Cycle *see* Bicycle
Cyclones **M** 22
Cynognathus **F** 29
Cyprus **X** 37
Czechoslovakia **X** 31, 32

▲ Cormorants are fish-eaters. The powerful beak, with curved tip, is used to catch fish. Cormorants are seabirds, although they can often be seen on rivers and reservoirs.

D

D'Arrest, Heinrich **N** 28
da Gama, Vasco *see* Gama, Vasco da
da Vinci, Leonardo *see* Vinci, Leonardo da
Dab **C** 40, 43
Daddy-long-legs **E** 38
Daedalus **Y** 44
Daffodils **A** 18, 19, 50; **G** 18–19
Daimler, Gottlieb **V** 30, 31; **W** 21
Daisies **A** 7, 31, 33
Daisy wheel printer **R** 26–27
Dalton, John **U** 12
Damselflies **E** 24, 26–27
Dandelion seeds **G** 22
Dandelions **A** 34
Daphnia **S** 23, 63
Darter **B** 24
Darwin, Charles **F** 9, 10–11; **G** 56
Data **Q** 9
Databases **Q** 39; **R** 47–48
David, Père **D** 32
Davy, Humphry **I** 26
de Coriolis, Gaspard *see* Coriolis, Gaspard de
de Dion Bouton **W** 22
de Dion tricycle **V** 16
de Forest, Lee *see* Forest, Lee de
de Fresne, Marion, *see* Fresne, Marion de
de la Cierva, Juan *see* Cierva, Juan de la
de Rivaz, Isaac *see* Rivaz, Isaac de
Dead Sea **J** 33
Deaf people, aids for **R** 50
Deafness **H** 38, 41
Death watch beetle **E** 48
Decapods **E** 53–54
Deciduous trees **A** 25
Decimal
  fractions **P** 13
  system **P** 10
Decisions **Q** 7, 9
Deep-sea angler **L** 44
Deep-sea squid **C** 22
Deer **D** 36, 50–51; **G** 60
Degrees **P** 18

◀ Deserts are not only sandy, as this photograph of the Canyon de Chelly in Arizona shows. Much of the erosion was caused by the river.

Deimos **N** 25
Deinonychus **F** 34
Deinosuchus **F** 38
Democritus **N** 35; **U** 12, 13
Dendrites **H** 32
Denmark **X** 28
Dentistry **I** 58–59
Deoxyribonucleic acid (DNA) **G** 49, 50; **K** 43
Depression **I** 60
Depressions **M** 15, 22–23
Descartes, René **P** 38
Deserts
  dusty **S** 25
  life in **M** 45
  locations **M** 44
  quartz grains **J** 21
  rats **D** 5
  Sahara **M** 44, 45
  sandstorms **M** 45
  temperatures **M** 45
Desmostylids **F** 56
Detective work **Y** 42–43
Detergents **U** 17
Devonian period **F** 14, 15
Dew
  advection **M** 8
  formation **M** 7–8
  point **M** 7
Diabetes **G** 53; **H** 29; **I** 40
Diagnosis **I** 9
Dialects **Q** 29
Diamonds **J** 19, 39; **S** 34; **Y** 27
Diaphragm **H** 19
Diatoms **L** 28; **S** 52–53
Diatryma **F** 48
Diaz, Bartholomew **W** 36
Dicotyledons **A** 19
Diesel, Rudolf **V** 32; **W** 24
Diesel engine **V** 28, 31–32, 50, 51
Diesel-electric locomotive **V** 25, 27
Diet **I** 59
Digestion
  birds **B** 12
  insects **E** 9
Digestive system, human **H** 11, 24–25, 50–51
Digigraph **R** 34
Digital recording **T** 46–47
Digital sounds **R** 36
Digitizers **Q** 45; **R** 32
Digitizing tablets **Q** 58
Dimetrodon **F** 25
Dingo **D** 29
Dinoflagellates **Y** 54
Dinornis maximus **F** 48
Dinosaurs
  age of **F** 26–28, 29; **G** 54
  armoured **F** 36
  bird-footed **F** 34–35
  flesh-eating **F** 30–31
  gazelles **F** 35
  giants **F** 32–33
  horned **F** 36–37
  mass death **F** 42–43
  plated **F** 36
Dion Bouton, de *see* de Dion Bouton
Dioptase **J** 13
Diplocaulus **F** 23
Diplodocus **F** 31, 32, 33
Diseases (volume **I**)
  congenital **I** 6, 35
  fungal **A** 11
  germ theory **I** 17–18
  inherited **I** 6, 35
  insects and **E** 16, 38
  names for **I** 6–7
  natural defences **I** 20–21
  plants **A** 11
Disk
  analogue board **Q** 54
  cantilever/loading mechanism **Q** 55

KEY TO TITLES OF VOLUMES

A Plants
B Birds
C Cold-blooded Animals
D Warm-blooded Animals
E Insects and their Relatives
F The World Before Man
G The Beginnings of Life
H The Human Body
I Disease and Medicine
J The World Beneath Us
K The Invisible World
L Seas and Oceans
M Weather and its Work
N Astronomy
P Mathematics
Q Computers: An Introduction
R Working with Computers
S Through the Microscope
T How Everyday Things Work
U Great Discoveries and Inventions
V How Does it Work?
W Communications and Transport
X The World Today
Y Projects

  drive **Q** 6, 54; **R** 25
  driver motor **Q** 55
  driving hub **Q** 54
  floppy **Q** 54, 55
  indicator **Q** 54
  protective envelope **Q** 55
  read/write head **Q** 55
  ribbon cable connector **Q** 55
  stepper motor and drive screw **Q** 55
  unit **R** 9
Displaying results **Q** 13
Dissection **I** 14
Divers
  birds **B** 20
  suits **L** 54–55

1600 TO 1800

1661 First paper money in Europe introduced in Sweden
1679 Binary arithmetic (in which all numbers are represented by the symbols 0 and 1) worked out by Gottfried von Leibnitz
1679 Pressure cooker invented by Denis Papin; originally intended to speed up cooking, it is also used in medicine for sterilizing instruments
1702 The first daily newspaper begins publication in England
1716 Sir Edmund Halley, the English astronomer, invents way of getting oxygen into diving bells
1752 Benjamin Franklin invents the lightning conductor
1760 George Dixon demonstrates that it is possible to light rooms with coal gas
1775 First modern water closet (water-flushed toilet) patented
1780 Modern screwdriver invented
1784 Bifocal spectacles introduced
1790 Dental drill invented by John Greenwood in the USA; it was driven by a spinning wheel
1791 Basic ideas of the metric system set out by the Paris Academy of Science; the first international system of measurement
1792 First modern (chemical) fire extinguisher introduced in Sweden
1794 Ball bearings used in a rotating windmill
1797 World's first parachute descent made by Jacques Garnier from a balloon at 1,000 metres

Dinosaurs dominated the earth for over 120 million years. This is a reconstruction of a plant-eating *Stegosaurus*.

KEY TO TITLES OF VOLUMES

A Plants
B Birds
C Cold-blooded Animals
D Warm-blooded Animals
E Insects and their Relatives
F The World Before Man
G The Beginnings of Life
H The Human Body
I Disease and Medicine
J The World Beneath Us
K The Invisible World
L Seas and Oceans
M Weather and its Work
N Astronomy
P Mathematics
Q Computers: An Introduction
R Working with Computers
S Through the Microscope
T How Everyday Things Work
U Great Discoveries and Inventions
V How Does it Work?
W Communications and Transport
X The World Today
Y Projects

tools **L** 56
Division **P** 12
Djibouti **X** 45
DNA *see* Deoxyribonucleic acid
Dodders **A** 6, 58
Dogs
  Cape hunting **D** 29
  carrying babies **B** 15
  domestic **D** 29, 36
  evolution **G** 51, 54, 56
  naming **G** 56
  pet **Y** 56
  sense of smell **Y** 42
  teeth **D** 29
  wild **D** 28–29
Dogwood **A** 25
Doldrums **M** 20
Dolphins **D** 60, 61; **L** 49; **W** 6
Dominican Republic **X** 19
Donkeys **D** 43
Dormice **C** 60, 61; **D** 30
Dot matrix printer **Q** 57; **R** 27
Doves **B** 38
Dragonflies
  breathing **E** 10
  colouring **E** 24
  giant **F** 22
  incomplete metamorphosis **C** 34
  larvae **E** 24–25
  life-cycle **E** 24–25
  mating **E** 25
  nymph **E** 25
  seeing **E** 13
  size **E** 24
Drake, Sir Francis **W** 38
Dravidosaurus **F** 36
Drawing, trick **Y** 8–9
Dreams **K** 46, 48, 49
Dredging **J** 44
Drilling **J** 36
Drinking **I** 60
Drizzle **M** 14
Dromedary **D** 48
Droughts
  causes **M** 26; **X** 57
  places suffering from **M** 26
Drug abuse **I** 60
Drugs **A** 48–49; **I** 23–24; **S** 37; **Y** 42
Du Pré, Jacqueline **I** 60
Duckbill **D** 12–13; **G** 44
Ducks **B** 6, 15, 26–28
Duckweeds **A** 18
Dugong **D** 11
Duikers **D** 58
Dummer, G W **Q** 27
Dung beetle **E** 18, 51
Dunlop, John **W** 18
Durer, Albrecht **P** 30
Dynamite **K** 34, 35; **V** 42

▲ Sir Francis Drake (*c* 1540–96) was one of the greatest English sailors. He sailed round the world in 1577–80.

▼ Husky dogs have extremely thick coats to help them withstand the cold of the Arctic winter.

E

KEY TO TITLES OF VOLUMES

A Plants
B Birds
C Cold-blooded Animals
D Warm-blooded Animals
E Insects and their Relatives
F The World Before Man
G The Beginnings of Life
H The Human Body
I Disease and Medicine
J The World Beneath Us
K The Invisible World
L Seas and Oceans
M Weather and its Work
N Astronomy
P Mathematics
Q Computers: An Introduction
R Working with Computers
S Through the Microscope
T How Everyday Things Work
U Great Discoveries and Inventions
V How Does it Work?
W Communications and Transport
X The World Today
Y Projects

Eagle **B** 12, 30–32
Early man *see Homo sapiens*
Ears
  animal **H** 39
  cochlea **H** 38
  eardrum **H** 38
  inner **H** 38
  middle **H** 38
  outer **H** 38
  oval window **H** 38
Earshells **C** 21
Earth
  Age for *see* Age for earth
  chemical elements **J** 8
  core **J** 8–10, 11, 12
  crust **J** 10
  diagrammatic section **J** 33
  distance form Sun **N** 12
  earth-moon system **N** 20–23
  exploration **J** 11–13
  forces **K** 32
  formation **J** 8–10
  geological structure **J** 11
  invisible **J** 8–10
  magnetic field **Y** 29
  main magnetic field **J** 12
  mantle **J** 8, 9, 12
  mapping **J** 11
  movement **Y** 22–23
  studying **J** 8–10
Earth-pig **D** 23
Earthquakes **J** 8, 9, 14, 39, 54–55
Earthworms **C** 14, 17, 34; **G** 11; **S** 23
East Africa **X** 45–46
East Germany **X** 30, 31
Easter Island **X** 51
Eastern Europe **X** 31–32
Echidna **D** 12, 13; **G** 44
Echinoderms **C** 24–25; **L** 35–37
Eckert, Presper **Q** 22
Eclipses
  lunar **N** 24
  solar **N** 24
Ecology notebook **Y** 60–61
Ecuador **X** 20, 22
Edaphosaurus **F** 25
Edentates **D** 22–23
Edible frog **C** 47
Edison, Thomas **W** 8
Edison phonograph **V** 48
Eel **C** 42
Eelworms **S** 23
Egg-laying mammals **D** 12–13
Eggs
  amphibians' **G** 37
  ants' **E** 46, 47
  aphids' **E** 30
  barnacles' **G** 26
  bees' **E** 40, 43; **G** 31
  beetles' **E** 51
  birds' **B** 6, 14–15; **G** 41, 43
  butterflies' **E** 34
  centipedes' **C** 27; **E** 59
  crabs' **E** 54
  crocodiles' **C** 50; **G** 39
  dragonflies' **E** 25
  earthworms' **C** 14; **G** 11
  eels' **G** 34
  egg-laying and **B** 6, 14–15
  fireworm **G** 27
  fish **G** 34, 36
  flies' **E** 38
  frogs' **C** 46; **G** 37, 38
  glow worms' **E** 22
  insect **G** 29
  internal fertilization **G** 27, 28
  ladybird **E** 19
  lice **E** 21
  lizard **G** 38, 39
  millipedes' **C** 27; **E** 58
  monotremes' **G** 44, 45
  mosquitos' **E** 17
  moths' **E** 14, 18, 34
  mussels' **G** 26
  newts' **C** 44; **G** 37
  orthinopeds' **F** 35
  Palolo worm **G** 27
  plants' **A** 12; **G** 16
  protozoan **G** 11
  reptiles' **G** 38, 39
  sea urchins' **G** 26
  spiders' **E** 57; **G** 30

▲ The eggs of the gorse shield bug are shaped like tiny barrels, with lids that the young push open when they hatch.

  termites' **E** 33; **G** 31
  toads' **G** 37, 38
  tortoise's **G** 39
  turtles' **G** 38, 39
  wasps' **E** 44, 45
Egypt **X** 9, 44
Egyptian cicada **E** 29
Egyptians **N** 6, 8; **W** 7; **Y** 24
Ehrlich, Paul **I** 23–24
Eiders **B** 26, 27
Einstein, Albert **N** 17; **P** 9, 42–43, 61; **U** 38
El Salvador **X** 18
Elands **D** 58; **G** 53
Elasmosaurus **F** 38
Eleatic School **P** 28
Electric buzzer **Y** 63
Electric car **W** 25
Electric motors **V** 27
Electric ray **C** 39
Electric telegraph **W** 8
Electrical charge **K** 29
Electrical energy **K** 24, 25
Electrical gadgets **Y** 30
Electrical generator **K** 26
Electricity
  atoms **V** 22–23
  cables **V** 23
  computers **Q** 21–22
  current **V** 24
  electrons **V** 22–23
  factories **V** 25, 36
  force **K** 27
  generating **V** 24, 36
  generating plant **V** 23
  generating station **V** 25
  generator **K** 26
  home **V** 25, 36
  information and **V** 27
  resistance **V** 24
  sub-station **V** 23
  testing **Y** 30–31
  voltage **V** 24
Electron beams **Q** 14
Electron microscope **S** 54–55, 60
Electron tube
  diodes **U** 56
  triodes **U** 56
Electronic clocks **R** 40
Electronic communications **W** 10–13
Electronic mail **W** 13
Electronic paintbox **R** 34–35
Electronics **U** 56–57
Electrons **K** 20, 28–29, 32; **Q** 7; **V** 22–23
Elements **U** 12
Elephants
  African **D** 16, 40
  hair **G** 44
  Indian **D** 40
  trunk **D** 40
  tusks and teeth **D** 40
Elevators
  construction **T** 16
  continuous lift **T** 16
  safety **T** 16
Elk **D** 50
Elms **A** 18
Embryo **G** 11, 21, 47, 53
Emperor dragonfly **E** 24
Emperor moth **E** 14
Emptiness **K** 6

Emu B 17
Encke's Comet N 32
Endocrine glands H 29
Energy
chemical K 25
conservation V 44–45
electrical K 24, 25
examples K 22
food and S 10
for life K 38–39
for the world J 45–46
force *see* Force
locked-up K 24
matter into K 30
movement- K 26
non-travelling K 24
quick K 38
slow K 38
sound K 26
stored K 26
travelling K 23, 24
what is K 24–25
Engine
diesel V 28, 31–32, 50, 51
four-stroke V 30, 31
internal combustion V 30–31, 32
jet V 9–11, 32, 40–41
meaning V 28
piston V 16–17
steam V 28–29
two-stroke V 31
Entertaining computers R 18–21
Enzymes K 42, 43
Eocene F 46
Eosuchians F 29
Epidemics I 29
Epilepsy I 35
Epiphytes A 22, 23
Equations P 14–15, 35
Equatorial Guinea X 45
Eratosthenes P 29
Ergot of rye S 31
Eriksson, Lief W 35
Eritreans X 46
Ermine D 34
Eryops F 23
Erythrosuchus F 28
Escalators
construction T 16
glass sided car T 17
sheave system T 17
Escher, Maurits P 19
Eskimos X 10, 11, 13, 14, 35, 55
Estimating P 13
Ethiopia X 40, 45, 46
Eucalyptus trees A 18
Euclid P 28, 35, 38
Euglena G 14; S 14, 20
Euler, Leonhard P 41
Euoplocephalus F 36
Euparkeria F 28, 29
Europe X 26–32, 55
European Economic Community X 28, 49
European lobster C 30
European pond tortoise C 51
Europeans X 10, 14, 15, 19, 22, 49, 51
Eusthenopteron F 22
Evaporation A 27
Everglades, Florida, USA J 24
Evergreen trees A 15
Evil K 58
Evolution A 10; C 34; G 54–60
Exact distortions Y 9
Examination I 8, 9
Exercise I 7
Experiments
colour Y 36–37
light Y 34–35
Expert systems Q 47
Exploration, great age of W 36–38
Explorers' mathematical aids P 35–37
Explosions J 11, 35
Explosive V 42–43
Extrasensory perception Y 40
Eyes
cells H 36
cones H 36
cornea H 36
insects E 12–13, 24
iris H 36
lenses H 36, 37
retina H 36, 37
rods H 36
sclera H 36
through the microscope S 43
working of H 36

▲ The flying bomb, or V1, used by the Germans against Britain during World War 2.

1800 TO 1900

1800 Modern porcelain false teeth developed
1816 René Laennec develops stethoscope from rolled up tube of paper used as an aid to hearing the heart beat
1818 James Blundell develops techniques of blood transfusion
1820 Charles Macintosh develops elastic
1824 Modern 'portland' cement introduced
1826 Patrick Bull develops corn reaping machine and exports it to USA to make up for labour shortages caused by the Civil War
1826 First 'friction' matches invented
1830 Food canning developed by Peter Durand
1830 Sewing machine introduced by Barthelemy Thimmonier and developed by Singer
1830 Edward Budding invents the lawn mower
1830 Plywood used for furniture by Michel Thonet
1834 Refrigeration developed by Jacob Perkins to meet the needs of the increasing Australia–London meat trade
1840 Postage stamp, and postal system, developed by Rowland Hill
1844 Paper patterns for home dressmakers given with women's magazines
1847 Charles Babbage develops ophthalmoscope to diagnose sight defects
1848 First use of khaki camouflage uniform for soldiers, by troops in India
1856 Introduction of modern synthetic dyes helps revolutionize textile industry
1860 Can opener introduced; the first cans had to be opened with a hammer and chisel
1860 Frederick Walton develops linoleum
1873 Joseph Glidden produces barbed wire by machine; soon widely used on ranches all over the USA
1876 M R Bissell introduces the carpet sweeper
1880 Blowlamp is developed in Sweden by Ludwig Holm and the Lindquist brothers
1884 Lewis Waterman introduces the first fountain pen
1889 H S Mills builds one-armed bandits – the first games machine to be made in quantity
1897 Charles Post makes Grape Nuts breakfast cereal and the Kellogg brothers introduce Corn Flakes
1899 Hermann Dreser in Germany introduces use of synthetic aspirin in medicine

F

F-111 fighter-bomber V 7, 10
F-47 Saber V 10
Fa 223 helicopter V 13
Fa-61 helicopter V 12–13
Fabrics S 25
Facsimile machine W 13
Factories R 54–60
Fahrenheit, Gabriel T 27
Fairlight R 36
Falcon B 10, 31, 32
Falcon, Robert Q 19
Fanworms C 15, 16
Faraday, Michael I 27
Farman, Henri W 48
Farming U 7, 8; X 8–9, 28, 29, 31, 33, 36, 56, 57
Farmyard poultry B 11
Fates K 56
Fats H 51
Feather-duster worm C 15
Feathers B 6, 7
Federated States of Micronesia X 51
Feeding
birds B12
insects E 8–9
Fell, James W 31
Fennec D 29
Fermat, Pierre de P 39
Ferns A 6, 13, 22, 23; G 16, 17
Fernwren B 52
Ferret D 34
Fertilization G 11, 13, 17, 19, 27–28, 34, 36, 38
see also Pollination
Fessenden, Reginald W 9
Fibres A 6; S 37; U 28–29
Fiddler crabs C 31; G 27–28
Field glasses Y 22
Fiji X 51
Filipinos X 15, 35
Film
animation Q 45–46
colour negatives T 32–33
control of amount of light reaching T 30–31
developing T 32
negative T 32
radioactivity detected by V 47
transparencies T 32
see also Motion picture
Finch-lark B 50
Finland X 28
Fir trees A 15
Fire U 6
Fire walkers K 60–61
Firebird Software R 38
Fireflies E 22
Fireworks V 18, 42
Fireworm G 27
First birds B 8
Fish
bony C 40–43; L 42–43
deep-sea L 44
defences L 42
finding food L 42–43
how they live L 42
pets Y 59
reproduction G 34, 36
senses Y 40, 41
sex changes G 13
Fishing
bottom netting L 58
catching fish L 58
drift netting L 58
fleets L 59
grounds L 58
plankton Y 54–55
Five to ten times bigger S 8–10
Flagellates C 8, 9; E 32; S 20
Flamingo B 25
Flanges U 6
Flatworms S 10, 13
Fleas E 16, 20, 21
Fleming, Alexander I 24
Fleming, John Ambrose W 10
Fleming, John U 56
Fliedner, Theodor I 50
Flies G 19; Y 40, 41
Flight, birds B 6, 9, 31
see also Airplane
Flight simulation Q 46; R 22–23
Flint J 22
Floods
causes M 26
disasters M 27
flash M 26–27
rivers M 26, 27
Flounder L 16
Flowcharts Q 35–36
Flowering plants A 6, 18–19, 22; F 46; Y 49, 50
Flowerpecker B 13
Flowers
arrangement A 31
colour A 30–31
garden A 50–51
importance of insects to E 18
male and female parts A 28–30
pollination G 18–19, 60
structure A 28
Fluke S 4, 47
Fluorspar J 62
Flying Y 44–45; see also Airplane, flight simulation
Flying fishes B 6; C 40, 42
Flying foxes D 18, 19
Flying lizard C 54
Flypapers, living A 55
Flywheel V 36
Focke, Heinrich V 12
Foetus G 47
Fog
arctic sea smoke M 8
formation M 8
freezing M 8
land M 8
radiation M 8
sea M 8
Foliage-gleaner B 47
Folk medicine A 48
Food
and energy S 10
chains A 9; L 28–9; X 59
digestion H 51; S 42
edentates D 22
effect on teeth I 58, 59
energy K 38
growing K 43

◀ Tropical ferns growing in profusion in a grotto at Kauai, Hawaii.

KEY TO TITLES OF VOLUMES

A Plants
B Birds
C Cold-blooded Animals
D Warm-blooded Animals
E Insects and their Relatives
F The World Before Man
G The Beginnings of Life
H The Human Body
I Disease and Medicine
J The World Beneath Us
K The Invisible World
L Seas and Oceans
M Weather and its Work
N Astronomy
P Mathematics
Q Computers: An Introduction
R Working with Computers
S Through the Microscope
T How Everyday Things Work
U Great Discoveries and Inventions
V How Does it Work?
W Communications and Transport
X The World Today
Y Projects

AFTER 1900

1901 First powdered instant coffee sold in USA
1901 Vaçuum cleaner powered by electric motor developed by Hubert Booth
1902 K C Gillette starts mass production of disposable razor blades for safety razors
1904 Caterpillar tracked vehicles developed by David Roberts in England
1904 Vacuum flask, used since 1892 for scientific purposes, introduced for use to keeps drinks hot on picnics etc
1910 First washing machine developed by Alva J Fisher
1913 Gideon Sundback of Sweden produces first practical zip fasteners
1914 First traffic lights used in USA; they were hand operated
1928 Jacob Schick introduces the first successful electric razor
1932 Parking meters introduced in the USA
1934 Catseye reflector developed by Percy Shaw comes into use on roads
1935 Juke-boxes first introduced in the USA
1937 Electric blankets go on sale in USA
1938 Plastic contact lenses developed
1938 Ladislao and Georg Biro develop the ballpoint pen that still bears their name
1944 Kidney machine developed by Willem Kolff comes into use
1948 Columbia Records introduce plastic long-playing records
1948 Microwave cookers developed
1953 RCA Records introduce the first music synthesizer
1955 Mark Grégoire uses a heat-resistant plastic to coat utensils, making the first non-stick pans
1958 Stereophonic recording comes into use
1963 First use of artificial heart to help a patient's circulation during an operation

KEY TO TITLES OF VOLUMES

A Plants
B Birds
C Cold-blooded Animals
D Warm-blooded Animals
E Insects and their Relatives
F The World Before Man
G The Beginnings of Life
H The Human Body
I Disease and Medicine
J The World Beneath Us
K The Invisible World
L Seas and Oceans
M Weather and its Work
N Astronomy
P Mathematics
Q Computers: An Introduction
R Working with Computers
S Through the Microscope
T How Everyday Things Work
U Great Discoveries and Inventions
V How Does it Work?
W Communications and Transport
X The World Today
Y Projects

microbes **S** 37
plants as **A** 6, 43–45
primates **D** 20
quantities to eat **I** 40–42
Foraminiferans **S** 16
Force
chemical **K** 34–35
earth **K** 32
electric **K** 32
energy and **K** 26–27
long-distance **K** 29
magnetic **K** 32
nuclear **K** 30
of attraction **K** 28
Ford, Henry **W** 22
Forecasting
weather *see* Weather forecasting
Forensic science **Y** 43
Forest, Lee de **V** 56
Forests
cold forest climates **M** 48–49
coniferous **A** 15
rain **A** 21–23, 60; **M** 38–39
redwood **A** 16
tropical **A** 21–23, 60
FORTH **Q** 48; **R** 14, 15
FORTRAN **Q** 48
Fossilized plants **A** 6, 18; **F** 6
Fossils
amphibians **C** 44
birds **B** 8
bones **F** 7
casts **F** 6, 13
collecting **Y** 20–21
coral animal **F** 7
fish **F** 18–19
formation **F** 6, 7
geology **J** 6, 11, 51, 56–57
guides to past **F** 8–9
in sedimentary rocks **J** 21
insects **E** 20, 21
living kingdoms **F** 11
meaning **J** 27–28
plants **F** 6
rock layers **F** 8–9
sea urchins **F** 9
shells **F** 6, 8
trees **J** 61
tuatara **C** 48
what they are **F** 6
Foucault, Jean **U** 43
Foxes **D** 28–29; **G** 56
Foxglove **A** 29
Foxglove seeds **G** 22
Fractions
algebra **P** 14
ordinary **P** 13
proper **P** 11
France **X** 19, 22, 28, 29–30, 40
Freezing
frost **M** 18
hail **M** 18
icicles **M** 19
method **K** 14
snow **M** 18–19
French Guiana **X** 22
French Polynesia **X** 51
Fresne, Marion de **C** 52
Freud, Dr Sigmund **I** 28; **K** 48
Freud, Martha **K** 48
Frilled lizard **C** 54
Frogfish **C** 40
Frogs **C** 46–47; **G** 11, 37; **S** 10
Frost
fern **M** 18
hoar **M** 18
rime **M** 18
Fruit bats **G** 23
Fruit fly **E** 38
Fruit pigeons **G** 23
Fruits **A** 34, 43, 44–45, 47
Fuel, rocket **V** 18–19
Fungal diseases of plants **A** 11
Fungi **A** 6, 10–11, 56–57 58–59; **S** 25, 26–27, 31, 37
Fur **D** 8
Fusulinids **F** 8

▼ Forest covers the sides of the Pindus mountains in Greece. Pines grow at the higher altitudes, giving way to beech on the lower slopes.

HIGHEST MOUNTAINS

These mountains are higher than 8000 m (26,247 ft). Some of them belong together because they are neighbouring peaks in a *massif* (a single large mountain block). All these peaks are in the Himalaya or Karakoram ranges, on the boundary of the Indian subcontinent and the central Asian landmass.

Mountain	Height (m)	(ft)
Mount Everest	8848	29,028
K2 (Chogori)	8610	28,250
Kangchenjunga	8597	28,208
Lhotse	8511	27,923
Yalung Kang Kangchenjunga West	8502	27,894
Kangchenjunga South Peak	8488	27,848
Makalu I	8481	27,824
Kangchenjunga Middle Peak	8475	27,806
Lhotse Shar	8383	27,504
Dhaulagiri I	8167	26,795
Manaslu I (Kutang I)	8156	26,760
Cho Oyu	8153	26,750
Nanga Parbat (Diamir)	8124	26,660
Annapurna I	8091	26,546
Gasherbrum I (Hidden Peak)	8068	26,470
Broad Peak I	8047	26,400
Shisha Pangma (Gosainthan)	8046	26,291
Gasherbrum II	8034	26,360
Annapurna East	8010	26,280
Makalu South-East	8010	26,280
Broad Peak Central	8000	26,247

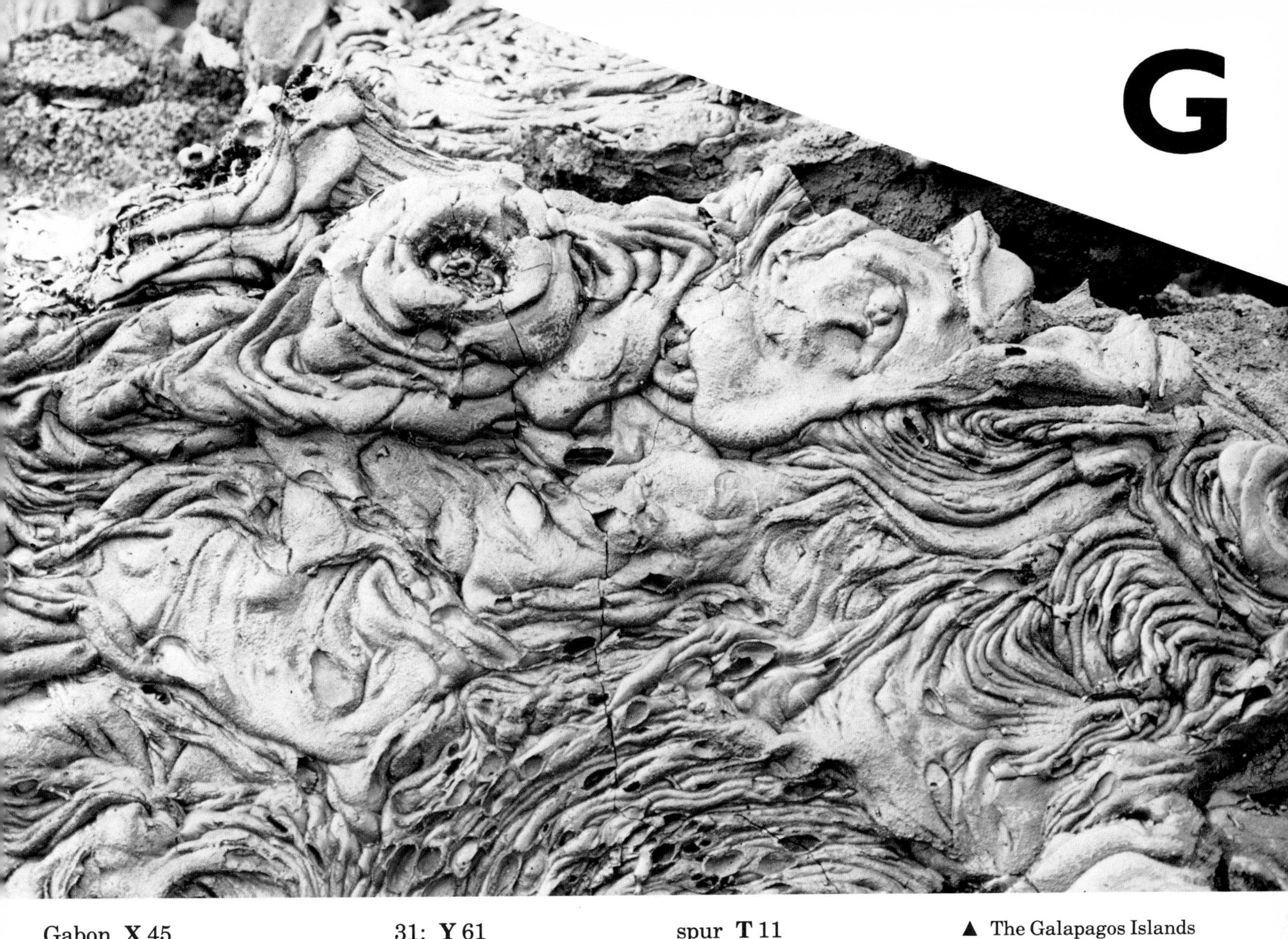

G

Gabon **X** 45
Gagarin, Yuri **V** 52
Galapagos cormorant **B** 24; **G** 59
Galapagos Islands **G** 59–60
Galapagos tortoise **C** 52
Galaxies **K** 6, 7; **N** 47, 48
Galilei, Galileo **N** 14, 16, 17, 23, 26, 35, 52–54; **P** 35; **U** 10, 11
Gall bladder **H** 25
Galle, Johann **N** 28
Gama, Vasco da **W** 36
Gambia **X** 45
Gamebirds **B** 33
Games **Q** 39–41; **R** 18–21, 40, 48–49
Games ports **Q** 17
Gametes **A** 12; **G** 11, 16
Gametophytes **A** 12
Gangrene **I** 23
Gannets **L** 51
Ganymede **N** 26
Garden flowers **A** 50–51; **Y** 49
Garden spider **C** 36
Garden tortoise **C** 51
Gardens **A** 50–51; **S** 28–31; **Y** 61
Gas turbine **V** 40–41
Gases
  breathed by humans **K** 10–11
  common **K** 8
  invisible **K** 8–9
  lightest **K** 8
  molecules **K** 12, 16
  pie chart **K** 8
  space rocket **K** 14–15
  structure **K** 6
  uncommon **K** 8
Gastropod molluscs **C** 20–21; **F** 16, 17; **L** 37; **Y** 20
Gastrotrichs **S** 23
Gaur **D** 56
Gavial **C** 50
Gazelle **D** 59
Gears
  bevel **T** 11
  bicycles **T** 7–8, 10
  helical **T** 10, 11
  herringbone **T** 11
  internal **T** 11
  planetary gear system **T** 11
  pumps **T** 15
  spur **T** 11
  uses **T** 10
  working **T** 10; **V** 17, 33, 35
  worm **T** 11
Geckos **C** 54
Geese **B** 15, 26, 28
Geiger, Johannes **V** 46
Geiger counter **J** 36; **P** 44; **V** 46–47
Geller, Uri **K** 59
General Dynamics Corporation **F** 16; **V** 7; **Y** 45
General practitioners **I** 49
General purpose register **Q** 12
General surgeons **I** 56
Genes **G** 49; **H** 54, 55; **I** 35; **S** 44; **U** 15
Genetics **G** 48–53; **H** 54
Gentians **A** 18
Gentoo penguin **B** 19
Geologists
  reading the past **J** 6
  work of **J** 6
  working with other scientists **J** 6

▲ The Galapagos Islands have striking geological formations as well as unique animals. These are old lava flows.

Geology
  further afield **J** 56
  in school **J** 56
  in the field **J** 56
  in the laboratory **J** 11
  in the landscape **J** 14–18
  in the museum **J** 56
  learning more about **J** 56–59
  meaning **J** 6
Geomagnetic dating **J** 13
Geomagnetism **J** 12
Geometry
  angles **P** 18
  classification **P** 16
  cubes **P** 16
  degrees **P** 18
  flat shapes **P** 16, 18
  measurement of shapes **P** 16
  patterns **P** 32–33
  philosophy and **P** 18

polygons **P** 16, 18
polyhedra **P** 18
quadrilaterals **P** 16, 18
solid shapes **P** 18
theorems **P** 38
triangles **P** 16
Geophones **J** 11
Geophysical research **J** 53
Geostationary orbit **W** 12
Geothermal heat **J** 53
Gerbert of Aquitaine **P** 31
Gerbils **D** 26
Gerenuk **D** 58
Germany **X** 28, 30, 31, 40, 49
Germs
theory **I** 17–18
types **I** 6
warfare against **I** 22–23
Gerrothorax **F** 23
Geysers **J** 47; **K** 33
Ghana **X** 45
Gharial **C** 50
Ghosts **K** 56, 57
Giant centipede **C** 27
Giant clam **C** 19
Giant octopus **C** 22
Giant panda **G** 60
Giant squid **C** 22
Giant tortoise **C** 51
Giant worms **S** 32
Giant's Causeway, Ireland **J** 17
Gibbons **G** 10
Gibraltar **X** 30
Gifford, Henry **W** 47
Gila monster **C** 54, 55
Gilbert, John **B** 49
Ginkgo tree **F** 21
Giraffes **D** 10, 54–55

KEY TO TITLES OF VOLUMES

A Plants
B Birds
C Cold-blooded Animals
D Warm-blooded Animals
E Insects and their Relatives
F The World Before Man
G The Beginnings of Life
H The Human Body
I Disease and Medicine
J The World Beneath Us
K The Invisible World
L Seas and Oceans
M Weather and its Work
N Astronomy
P Mathematics
Q Computers: An Introduction
R Working with Computers
S Through the Microscope
T How Everyday Things Work
U Great Discoveries and Inventions
V How Does it Work?
W Communications and Transport
X The World Today
Y Projects

Gizzard **B** 12
Glaciers **L** 14–15; **M** 58
Glands
blood passing through **H** 20
functions **H** 28
sweat **H** 28
Glass **K** 16
Glass snakes **C** 54
Glassblowers **U** 33
Glassmaking
early **U** 32–33
float process **U** 33
modern **U** 33
Glenn, John **V** 52
Gliden, Christopher **Q** 11
Glider **V** 6, 14
Globe **J** 10
Glowworms **E** 22
Glutton **D** 34–35
Gnawing animals *see* Rodents
Goat-antelopes **D** 53, 57
Goats **D** 52–53
Goatsuckers **B** 42
Goddard, Robert H **U** 51
Gogh, Vincent van *see* Van Gogh, Vincent
Gold **K** 18; **Y** 19
Goldfish **G** 36; **Y** 59
Gondwanaland **F** 25, 27, 28
Goodyear, Charles **U** 34
Goosander **B** 12
Goosegrass **G** 23
Gorilla **D** 20
Gorse seeds **G** 22
Graetz, Martin **R** 18
Grand Canyon **J** 6, 7, 29
Grand Prix racing **R** 21
Granite **J** 19, 22
Grapes **A** 46; **Y** 46
Graphics
animation **R** 30–31
block **R** 29, 30, 32
electronic paintbox **R** 34–35
games **R** 18, 21
light pens **R** 34
made easy **R** 32–35
on paper **R** 35
simple **R** 29
sprites **R** 30, 31
tablets **R** 32
uses **R** 28
Graphs
different types **P** 51–52
misusing **P** 51
pie diagram **P** 51–52
three dimensions **P** 52
uses **P** 50–51
Graptolites **F** 15
Grass snakes **C** 57; **G** 38–39
Grasses **A** 19, 28, 33, 36–39
Grasshoppers **C** 34; **E** 6, 8, 9, 10, 13, 16, 28
Grasslands
temperate **M** 43
tropical **M** 42–43
Gravimeters **J** 12–13, 36
Gravitation **U** 38–39; **V** 52
Gravity **J** 12–13; **K** 28, 29, 30, 58; **N** 16–17; **P** 40
Great Bear (Ursa Major) **N** 6, 7, 36, 37; **Y** 22
Great diving beetle **E** 7, 48
Great Plague **I** 13
Great white shark **C** 38
Grebes **B** 21
Greece **X** 27, 28, 31, 49
Greek alphabet **W** 7
Greek cicada **E** 29
Green turtle **C** 52
Greenbottle **S** 26
Greenfly **E** 8
Greenland **L** 15; **X** 13
Greenwich Mean Time **P** 46
Greenwich Observatory **P** 45, 46
Gristle fish *see* Cartilaginous fish
Grouse **B** 33
Grubs **C** 34; **E** 15; **G** 33
Guam **X** 51
Guanaco **D** 49
Guatemala **X** 18
Guillemot **B** 13
Guinea **X** 45
Guinea pigs **D** 26–27
Guinea-Bissau **X** 45
Gulf of Genoa, Italy **J** 51
Gulf Stream **L** 12
Gulls **B** 31, 37
Gulpers **L** 44
Gum disease **I** 59
Gunpowder **V** 42
Guns **Q** 15
Gutenberg, Johann **U** 10, 24; **W** 7
Guyana **X** 22
Gymnosperms **A** 15; **F** 20, 21, 28
Gyrocompass **U** 43
Gyroscope
inertia **U** 43
precession **U** 43
spinning **U** 43
what it is **U** 43

▲ Grand Prix motor racing tests the skills of the world's best racing drivers and the design and engineering of their vehicles.

► Helicopters are immensely versatile. They can be used to reach places quite inaccessible to any other vehicle.

H
ARMY

KEY TO TITLES OF VOLUMES

A Plants
B Birds
C Cold-blooded Animals
D Warm-blooded Animals
E Insects and their Relatives
F The World Before Man
G The Beginnings of Life
H The Human Body
I Disease and Medicine
J The World Beneath Us
K The Invisible World
L Seas and Oceans
M Weather and its Work
N Astronomy
P Mathematics
Q Computers: An Introduction
R Working with Computers
S Through the Microscope
T How Everyday Things Work
U Great Discoveries and Inventions
V How Does it Work?
W Communications and Transport
X The World Today
Y Projects

H-bomb **K** 30, 31, 32
Habitats **Y** 60, 61
Hadrosaurids **F** 34
Haemoglobin **H** 20
Hahn, Otto **U** 54
Hailstones **M** 18
Hair
different kinds **D** 8
follicles **D** 8; **H** 46; **S** 38, 41
insulation **D** 8
Hairybacks **S** 23
Haiti **X** 19
Half-parasites **A** 58
Hall, Chester Moor **N** 55
Hall, Samuel **W** 39
Halley, Edmond **N** 16, 32
Halley's Comet **N** 10, 32
Hallucinations **H** 57
Halteres **E** 38
Hammerhead shark **C** 38
Hamsters **D** 26; **Y** 56
Hancock, Thomas **W** 18
Handicapped people, help for **R** 50–51
Hang gliders **V** 14
Hard copy **R** 26
Hardware **R** 9
Hardwood **A** 14
Hares
coat **D** 25
Jack rabbit **D** 24
marsh **D** 25
Nelson **D** 25
Harrison, John **W** 42, 43
Harvest mice **D** 26
Harvey, William **H** 10–11, 20; **I** 15
Harz mountains, Germany **Y** 19
Hatchetfish **C** 42
Haunting **K** 57
Hawaii **X** 15, 51
Hawk **B** 10, 31
Hawk moths **E** 34
Hawker Siddeley Harrier **V** 9
Head louse **E** 21
Health
talking about **I** 6–7
Hearing
birds **B** 11
human **H** 38–39
insects **E** 13
testing **Y** 40
Heart
attack **I** 44
beat **B** 9
blood circulation **I** 15
chambers **H** 18
defects **I** 36
disease **I** 43–44
function **H** 18
insects' **E** 10
muscle **H** 31; **I** 43
pacemaker **I** 44
partnership with lungs **H** 19
rhythm **H** 18
surgery **I** 56
tissues **S** 42
transplants **I** 57, 58, 60
Heat **H** 35, 60; **K** 22, 24, 33
Heat rays **S** 6–7
Heaviest substances **K** 18
Hectocotylus **G** 28
Hedgehog **C** 61; **D** 16, 17
Heinkel, Ernst **W** 53
Heinkel He 178 **V** 9
Hektor **N** 30
Helicopter
agility **V** 12
first **V** 12; **W** 56–57
lift **V** 13
modern **V** 12–13
passenger-carrying **V** 12
rotors **V** 13
Heliozoans **S** 16
Helium **K** 8, 9
Hematite **J** 37
Hemicyclaspis **F** 19
Hemoglobin **K** 37
Hemophilia **I** 35–36
Henodus **F** 38
Henry, Prince of Portugal **W** 36
Hepatits **S** 59
Herbal preparations **I** 14
Herbivores **F** 52–54
Hercules beetle **Y** 58
Herding **U** 7
Heredity **G** 48, 57; **H** 54–55; **K** 43
Hermann's tortoise **C** 51
Hermaphrodites **G** 11
Hermit crabs **C** 32; **E** 53–54
Heroin **Y** 42
Herons **B** 25
Herrings **C** 40; **G** 34
Herschel, Caroline **N** 46
Herschel, John **N** 46
Herschel, William **N** 28, 46, 55
Hertz, Heinrich **U** 52
Hesperornis **F** 47
Hi-fi connector **Q** 11
Hibernation **C** 7, 58; **D** 6, 30, 62
Hieroglyphics **W** 6, 7
High speed trains **W** 30
High tension circuits **Q** 15
Hindus **X** 38, 39
Hipparchus **N** 40
Hippocrates **I** 11–12
Hippopotamus **D** 46
History of plants **A** 6
Hogweed **A** 31
Holland, John **V** 50
Hollerith, Herman **Q** 19–20, 22
Holly leaves **A** 24, 25
Holly tree **G** 18
Holmes, Sherlock **Y** 42
Holocene times **F** 61
Hologram
making **U** 60, **Y** 7
uses **U** 60–61
viewing **U** 60
Holography **U** 60–61; **Y** 7
Home computers **Q** 8–9, 11; **R** 9, 19, 38
Homes **S** 25–27
Homo erectus **X** 7
Homo habilis **X** 6–7
Homo sapiens **J** 30; **X** 7

▼ Horses have been used for sport for thousands of years. This print shows a game of polo in 8th-century China.

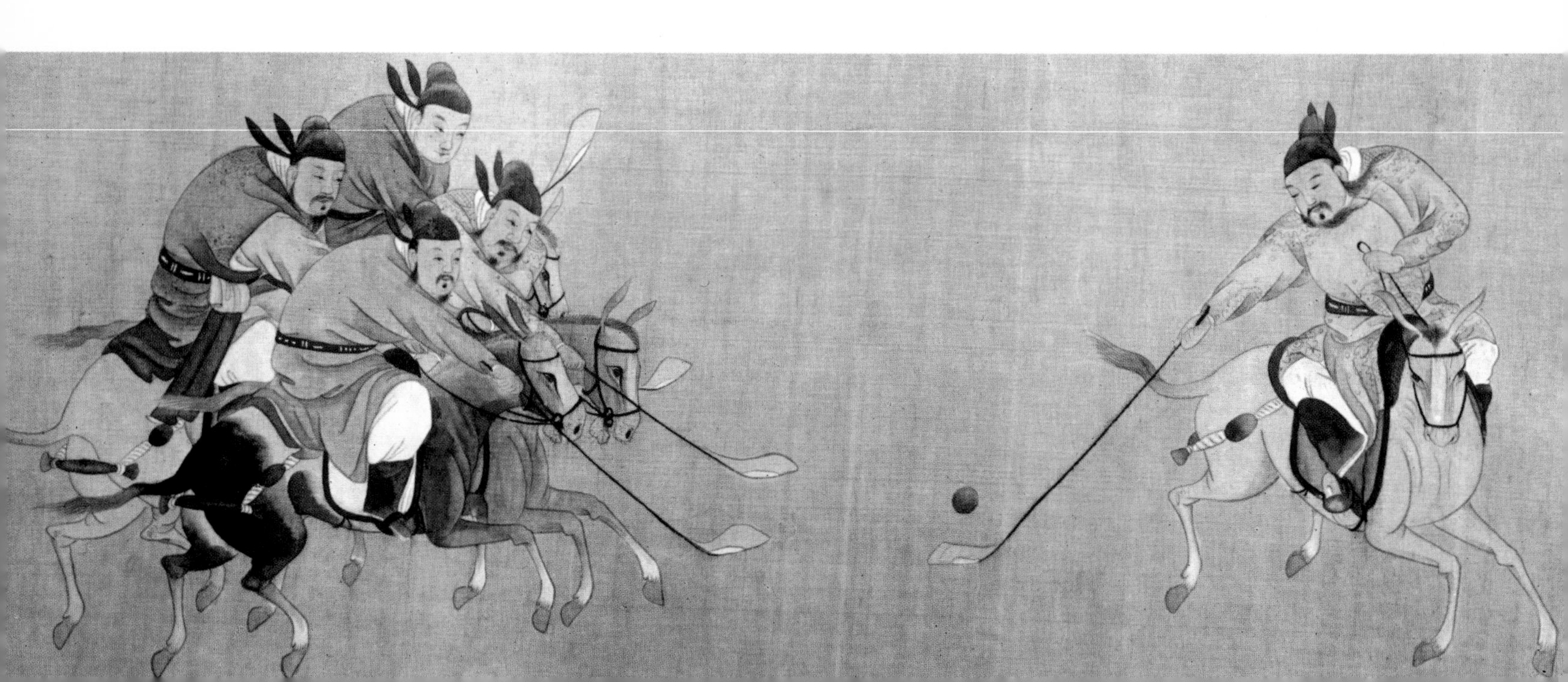

Homo sapiens neanderthalensis **X** 8
Honduras **X** 18
Honey-guides **B** 40
Honeybees **A** 29, 32; **C** 33; **E** 19, 40–42
Honeydew **E** 31
Hong Kong **X** 36
Honshu, Japan **J** 15
Hooke, Robert **U** 15
Hoopoes **B** 45
Hormones **H** 20, 28–29, 53; **I** 40
Hornbills **B** 45
Horned frog **C** 47
Horned lizard **C** 55
Horned toad **C** 54
Hornets **E** 45
Horse latitudes **M** 20
Horse-power **V** 29
Horsedrawn transport **W** 19, 26
Horsefly **E** 7
Horses **D** 42–43; **G** 55
Horseshoe bat **C** 61
Hospitals
  day care services **I** 52
  different parts **I** 52
  isolation **I** 52
  laboratories **I** 54
  maternity department **I** 52
  wards **I** 52
Hottentots **X** 10, 44
House martins **B** 51
House plants **Y** 50
Housefly **E** 38
Hoverflies **E** 38
Hoyas **A** 19
Hubble, Edwin Powell **N** 47, 48, 49
Hughes, Professor David **W** 8
Humason, Milton **N** 47, 48
Humason's Comet **N** 32
Humidity
  dew **M** 7–8
  fog **M** 8
  forecasting **M** 31
  meaning **M** 7
  mist **M** 8
  relative **M** 7, 31
Hummingbirds **A** 32; **B** 9, 12, 15, 44, 55; **C** 58, 59; **G** 19
Hungary **X** 31, 32
Hunting **U** 6, 7
Hurricanes
  development **M** 24–25
  distance travelled **M** 25
  winds **M** 24
  worst-hit coasts **M** 25
Hutton, James **F** 8
Huygens, Christiaan **N** 54, 55
Hyacinth plant **Y** 49
Hyatt, Isaiah **U** 29, 30
Hyatt, John Wesley **U** 30
Hybridization **G** 51, 53
Hydras **C** 13; **S** 13
Hydrocarbons **J** 30, 45
Hydroelectric power stations **U** 44, 45
Hydroelectricity **U** 44–45
Hydrofoil **W** 40
Hydrogen **K** 8, 9, 18; **U** 12
Hydrogen bomb **V** 42
Hydrological cycle **J** 48
Hyenas **D** 35
Hygrograph **M** 31
Hypnotism **K** 59
Hypsilophodontids **F** 34, 35
Hyracotherium **F** 46
Hyrax **D** 41

▲ This hybrid is the result of crossing a horse and a zebra. Hybrids, which rarely occur naturally, are unable to breed and have young.

THE WORLD'S OCEANS

These are the Earth's oceans or seas that are more than 1 million sq km (620,000 sq mi) in area. You may find that different reference sources give slightly different sizes, since seas have only roughly defined limits. The seas cover five-sevenths of the Earth's surface.

	sq km	sq mi
Pacific Ocean	165,380,000	63,854,000
Atlantic Ocean	82,220,000	31,745,000
Indian Ocean	73,480,000	28,371,000
Arctic Ocean	14,060,000	5,429,000
Mediterranean Sea	2,500,000	965,000
South China Sea	2,320,000	896,000
Bering Sea	2,270,000	876,000
Caribbean Sea	1,940,000	749,000
Gulf of Mexico	1,540,000	595,000
Sea of Okhotsk	1,530,000	591,000
East China Sea	1,250,000	483,000
Hudson Bay	1,230,000	475,000
Sea of Japan	1,010,000	390,000

I

▲ The icefields of the Arctic and the Antarctic vary little throughout the year. Here a glacier in the Antarctic is lit by the pale light of the midsummer Midnight Sun.

Iatrogenic diseases **I** 60
Ibex **D** 52
Icarus **N** 30; **Y** 44
Ice
Age **F** 58–61; **M** 58
changing landscape **J** 24
clouds **M** 10
debris dumped by **M** 58–59
effects on land **M** 58
melting **K** 16
see also Sea ice
Icebreakers **W** 41
Icefish **C** 6–7
Iceland **X** 28
Ichthyornis **F** 47–48
Ichthyosaurs **F** 39
Icicles **M** 19
Iguanas **C** 55
Iguanodon **F** 34, 35
Illness, types of **I** 6
Illusions, optical **Y** 38–39
Immunity **I** 19–21
Inca empire **U** 8; **X** 20–21, 22
Incas of Peru **D** 26
Inclined plane **V** 35
Incomplete metamorphosis **E** 14, 15; **G** 29, 33
India **X** 10, 38, 56
Indian Ocean **C** 19; **J** 33; **L** 7, 8, 16, 24
Indian moon moth **E** 36
Indonesia **X** 39
Industrial blower **Y** 31
Industrial revolution **X** 9, 27, 29
Industry **S** 37
Infection **H** 20; **I** 22; **S** 48
Inflammations **I** 19, 20
Influenza **I** 30; **S** 59
Information services **R** 44, 47–48, 50
Input unit **Q** 9
Input/output **Q** 35
Insect-eating bat **C** 61
Insectivores **D** 16–17
Insects (volume **E**)
as pets **Y** 58
basic body design **C** 33–34
biological control **E** 18–19
communication **W** 6
early **F** 22
flight **B** 6
larvae **C** 17
life history **C** 34; **E** 14–15; **G** 29–33
meaning **E** 6
plants which eat **A** 54, 55
pollination **A** 32, 33; **G** 19
reproduction **G** 13, 28
species **G** 55
through the microscope **S** 9, 29
useful **E** 18–19
Inside computers **R** 10–11
Instinct **K** 54–55
Institute of Geological Sciences **Y** 21
Instructions, vital **R** 14–15
Insulation **D** 8
Insulin **H** 29
Integers **P** 10
Integrated circuits *see* Chips
Intelligence **K** 50–53
Intelsat system **W** 12
Interferon **I** 31–32
Internal combustion engine **V** 30–31, 32; **W** 20–21
Internal fertilization **G** 27–28, 36, 38
International Ice Patrol **L** 15
Intestines, human **H** 20, 24, 25, 26
Inuit **X** 14, 5
Inventions mathematical **P** 38–39
Invertebrates
early **F** 22
simple **C** 8–17
Ionosphere **W** 12
Iran **X** 37, 38
Irano-Afghans **X** 35
Iraq **X** 9, 37, 38
Ireland **X** 28, 29
Iron **J** 21, 32; **K** 32; **U** 11; **Y** 28, 32
Iron Age **J** 30
Iron ores **J** 30
Ironstone nodules **Y** 21
Irrigation **X** 9, 20
Irwin, Jim **V** 55
Islam **X** 33, 35, 44
Islands
continental **L** 24
effects of isolation on wildlife **G** 59
oceanic **L** 24
plants on **A** 60–61
Isolation **G** 58–59, 60
Isopods **L** 34
Israel **X** 37
Italy **X** 28, 30–31, 49
Ivory Coast **X** 45

Jack **V** 33, 35
Jackals **D** 29
Jackdaw **B** 61
Jacquard, Joseph **Q** 19, 20
Jaguar **D** 36, 37
Jamaica **X** 19
Jansky, Karl **N** 56
Janssen, Hans **U** 15
Janssen, Zacharias **U** 15
Japan **X** 36, 37
Japanese **X** 11, 15, 35, 49
Japanese crab **C** 31
Japanese sound system **V** 48
Javelin rocket **V** 19
Jays **B** 61; **G** 23
Jeffires, Dr John **W** 47
Jellinek, Mercedes **W** 22
Jellyfish **C** 12; **G** 54; **L** 30; **S** 12, 13
Jentazy, Camille **W** 25
Jenner, Edward **I** 17–18
Jet age **W** 53–55
Jet engine **V** 9–11, 32, 40–41
Jews **X** 10, 22, 33, 35, 37
Joints
  ball and socket **H** 16
  hip **H** 17
  mobility **H** 16
Jordan **X** 38
Joystick **Q** 6, 56; **R** 22
Joystick connector **Q** 10
Jump jets **W** 57
Jung, Carl **I** 38
Jungles *see* Forests, rain
Juno **N** 30
Jupiter
  cloud belts **N** 26
  discovery **N** 10
  moons **N** 26; **V** 57
  orbit **N** 11
  probes **V** 56
  revolving of **N** 26
  rotation **N** 26
  size **N** 26
Jurassic times **F** 28

▲ Jungles, the popular name for the thickest tropical rainforests, have been cleared in many parts of the world. Here, in Hawaii, are some of the few surviving areas of jungle.

LONGEST TUNNELS

These are the world's longest road and rail tunnels. Even longer tunnels than these have been built to carry water.

km	miles		
53·9	33·49	Seikan (rail) 1972–82	Tsugaru Channel, Japan
30·7	19·07	Moscow Metro 1979	Belyaevo to Medvedkovo, Moscow, USSR
27·84	17·30	Northern Line (Tube) 1939	East Finchley-Morden, London
22·17	13·78	Oshimizu 1981	Honshū, Japan
19·82	12·31	Simplon II (rail) 1918–22	Brigue, Switzerland-Iselle, Italy
19·80	12·30	Simplon I (rail) 1898–1906	Brigue, Switzerland-Iselle, Italy
18·68	11·61	Shin-Kanmon (rail) 1974	Kanmon Strait, Japan
18·49	11·49	Great Apennine (rail) 1923–34	Vernio, Italy
16·32	10·14	St Gotthard (road) 1971–80	Göschenen-Airolo, Switzerland
16·0	10·00	Rokko (rail) 1972	Japan
16·0	9·94	Hong Kong Subway 1975–80	Hong Kong
15·8	9·85	Henderson (rail) 1975	Rocky Mts, Colorado, USA
14·9	9·26	St Gotthard (rail) 1872–82	Göschenen-Airolo, Switzerland
14·5	9·03	Lötschberg (rail) 1906–13	Kandersteg-Goppenstein, Switzerland
14·0	8·7	Arlberg (road) 1978	Langen-St Anton, Austria
13·85	8·61	Hokkuriku (rail) 1957–62	Tsuruga-Imajo-Japan
13·6	8·5	Mont Cenis (rail extension) 1857–81	Modane, France-Bardonecchia, Italy
13·35	8·3	Shin-shimizu (rail) 1967	Japan
13·03	8·1	Aki (rail) 1975	Japan

K

▲ Kite flying with a difference! But you'll need plenty of wind to get lift-off for a kite and a passenger.

KEY TO TITLES OF VOLUMES

A Plants
B Birds
C Cold-blooded Animals
D Warm-blooded Animals
E Insects and their Relatives
F The World Before Man
G The Beginnings of Life
H The Human Body
I Disease and Medicine
J The World Beneath Us
K The Invisible World
L Seas and Oceans
M Weather and its Work
N Astronomy
P Mathematics
Q Computers: An Introduction
R Working with Computers
S Through the Microscope
T How Everyday Things Work
U Great Discoveries and Inventions
V How Does it Work?
W Communications and Transport
X The World Today
Y Projects

Kallima butterfly **E** 36
Kampuchea **X** 39
Kändler, J **U** 20
Kangaroo
  evolution **G** 60
  joey **D** 14
  national emblem of Australia **B** 17
  rat **D** 14
  red **D** 14
Kanimbla mountains, Australia **F**15
Kannemeyeria **F** 29
Kant, Immanuel **N** 33
Katydid **E** 29
Kawasaki motorcycle **V** 17
Kay, John **U** 27
Kentrosaurus **F** 3, 36
Kenya **X** 40, 46
Kepler, Johannes **N** 16, 32, 53, 54
Kepler's Laws **N** 16
Kestrels **B** 11, 32
Keyboard **Q** 9, 10, 11; **R** 51
Keyboard connector **Q** 16
Keys
  ancient **T** 19
Kidney
  artificial **H** 27
  collecting tubes **H** 27
  functions **H** 25
  injury **H** 25, 27
  tissue **S** 42
  transplants **H** 27; **I** 57–58
Kilby, Jack **Q** 27
Kilobytes **Q** 26
Kinetic energy **V** 45
King penguins **B** 18, 19
King protea **A** 30
Kingfishers **B** 45
Kinkajou **D** 33
Kiribati **X** 51
Kitchens **S** 25
Kites
  birds **B** 31
  working **V** 6
Kiwis **B** 11, 18
Klipspringer **D** 58
Knitting machine **U** 26
Knowledge-based program **Y** 14–15
Koala **D** 14, 15
Koala-pad **R** 32–33
Koch, Robert **I** 17, 23, 24
Komodo dragon **C** 54, 55
Kookaburra **B** 45
Korea **X** 36
Krakatoa **K** 32
Krill **D** 60
Kronosaurus **F** 39
Kurds **X** 38
Kuwait **X** 37

▲ Lion cubs hiding in bushes – perhaps waiting for the female to bring them food.

Labidosaurus **F** 24
Lablace, Pierre **N** 33
Laboratories **I** 54; **J** 11
Labyrinthodonts **F** 23
la Cierva, Juan de *see* Cierva, Juan de la
Ladybird **E** 18, 19, 30, 31, 50, 51; **S** 6
Lagoon **L** 22, 25
Lammergeier **B** 29
Land
crabs **G** 28
ice effects on **M** 58
snails **G** 28
wind effects on **M** 60
Landscape
changing **J** 23–24
geology in **J** 14–18
junmbled **J** 14–15
Language
high-level **Q** 13, 48
low-level **Q** 48
peoples and **X** 13, 14, 18, 19, 22, 27–28, 29–30, 32, 36, 38–39, 40, 45, 46
programming **Q** 28–29
using **W** 6–7
Lanzarote, Canary Islands **Y** 24
Laos **X** 39
Lapps **X** 28, 55
Larches **A** 15
Larks **B** 50
Larvae **C** 45; **E** 9, 10, 14, 15; **G** 11, 27, 29, 33; **L** 28
Larynx **H** 40; **W** 6
Lasers
beams **Y** 16
gas **U** 58, 59
light **U** 58
solid **U** 58
story of **U** 58
Latin America **X** 14, 19, 22
Laughing gas **I** 26
Laughing jackass **B** 45
Laurasia **F** 27
Lawyer's wig fungi **A** 11
Le Verrier, Urbain **N** 28
Lead **Y** 32
Leaf-cutter ants **E** 46
Leafbeetles **E** 50
Learning processes **P** 60–
Leathery turtle **C** 52–53, 62
Leaves
gas exchange **A** 25
other uses **A** 25
stoma **A** 24
structure **A** 24–25
Lebanon **X** 38
Leeches **C** 17
Leewenhoek, Anton van **H** 8; **I** 15; **S** 52–3; **U** 15
Legumes **A** 42
Leibnitz, Gottfried von **P** 38–39, 41; **Q** 20
Lemurs **D** 20
Lenoir, Jean **W** 20
Lenses
binoculars **T** 34
blooming **T** 35
camera **T** 33, 34
construction **T** 33
convex **T** 33–34
magnifying glass **T** 33, 35
position of **T** 34
spectacles **T** 35
Lentinellus fungi **A** 11
Leonids **N** 31
Leopard **D** 36, 37
Lepidoptera **E** 34
Lepospondyls **F** 23
Lesage, Georges **Q** 21
Lesotho **X** 46
Leukemias **I** 46
Levassor, Emile **W** 21, 22
Lever **V** 34, 35
Lianes **A** 22
Liberia **X** 40, 45
Libya **X** 44
Lice **E** 21
Lichens **A** 56–57
Liechtenstein **X** 30
Life
beginnings **F** 12–13; (volume **G**)
below sea **F** 14–15
cycles **A** 34–35; **G** 11
history **E** 14–15

KEY TO TITLES OF VOLUMES

A Plants
B Birds
C Cold-blooded Animals
D Warm-blooded Animals
E Insects and their Relatives
F The World Before Man
G The Beginnings of Life
H The Human Body
I Disease and Medicine
J The World Beneath Us
K The Invisible World
L Seas and Oceans
M Weather and its Work
N Astronomy
P Mathematics
Q Computers: An Introduction
R Working with Computers
S Through the Microscope
T How Everyday Things Work
U Great Discoveries and Inventions
V How Does it Work?
W Communications and Transport
X The World Today
Y Projects

▼ Lyrebirds get their name from their tail which is shaped like the ancient musical instrument. These birds live in Australia.

in water **F** 38–39
Lift *see* Elevator
Light
energy **K** 22
experiments with **Y** 34–35
pens **Q** 59; **R** 34
rays **S** 6, 7
Lightest substances **K** 18
Lighthouses **L** 53
Lightning
electrical charges **V** 22
flashes **Y** 30
forked **M** 17
meaning **M** 16–17
risks from **M** 17
sheet **M** 16
Lilienthal, Otto **W** 48
Lilies **A** 19; **G** 20
Lily flower **S** 8
Limpets **C** 21; **G** 13, 27
Lindbergh, Charles A **V** 8
Linnaeus, Carl Linné **G** 56
Lion **D** 36, 37
Lippershey, Hans **N** 52
Lips **E** 8
Liquids
capillary action **K** 13
curious habits **K**12
molecules **K** 12, 16
structure **K** 6
surface-binding force **K** 12
what are **K** 12–13
LISP **R** 16
Lister, Joseph **I** 22–23
Liver
bile **H** 25
fluke **I** 33
functions **H** 25
tissues **S** 42
Liverworts **A** 12; **G** 16, 17
Living light **E** 22
Lizards **C** 6, 7, 54, 55; **F** 31; **G** 13, 38, 39; **Y** 58
Llamas **D** 49
Lobachevski, Nikolai **P** 42
Lobsters **C** 28, 30, 34; **E** 52, 53, 55; **L** 34
Locks
cylinder rim **T** 19
keys **T** 19
mortise **T** 19
yale **T** 19
Locomotives **K** 25; **W** 27–28, 30
Locusts **E** 16, 17
Loewy, M **N** 52
Logical operation **Q** 31
LOGO **Q** 48–49; **R** 15
Loons **B** 20
Looping **Q** 32–34
Lories **B** 39
Lorikeets **B** 39
Lovebirds **B** 39
Lovelace, Ada, Countess **Q** 21
Luciferin **E** 22
Lugworm **C** 16
Luminous insects **E** 22
Lunaspis **F** 19
Lungfish **F** 19; **G** 37
Lungs
alveoli **H** 19
birds **B** 9
function **H** 18
partnership with heart **H** 19
rhythm **H** 19
tissue **H** 18–19; **S** 42
Luxembourg **X** 28
Lymph
functions **H** 22
glands **H** 22
nodes **H** 22, 23
system **H** 22–23
vessels **H** 22, 23
Lymphocytes **H** 22
Lymphomas **I** 46
Lynx **D** 36, 37
Lyrebirds **B** 8

► Microscopes, especially powerful electron microscopes, can certainly alter our view of the world. This is not a huge pipe, but the tip of a hypodermic needle!

M

Macao **X** 36
Macaws **B** 39
Macchu Picchu **U** 8
Machine intelligence **Q** 47
Madagascar **X** 46
Magellan, Ferdinand **B** 60; **N** 47
Maggots **E** 15; **G** 33
Magic
boxes **P** 31
mathematics as **P** 30–31, 34
Magnesium **K** 37
Magnetic compass **Y** 29, 32
Magnetism **K** 32; **Q** 7; **Y** 28–29
Magnetometer **J** 36
Magnifying glass **T** 33, 35
Magnolia **A** 18; **F** 46
Magpie **B** 61
Maidenhair tree **A** 15
Main circuit board **Q** 15
Mainframe computers **P** 55, 59; **Q** 8, 9; **R** 18; **Y** 14
Maize **A** 37
Malaria **E** 17; **I** 34; **S** 48
Malawi **X** 46
Malaysia **A** 21; **X** 39
Maleo fowl **B** 34
Mali **X** 44
Mallee fowl **B** 34
Malta **X** 31
Mammals (volume **D**)
air **F** 55
bats *see* Bats
characteristics **D** 7
early **F** 49–50
egg-laying **D** 12–13
hair and fur **D** 8–9
hibernation **C** 61

KEY TO TITLES OF VOLUMES

A Plants
B Birds
C Cold-blooded Animals
D Warm-blooded Animals
E Insects and their Relatives
F The World Before Man
G The Beginnings of Life
H The Human Body
I Disease and Medicine
J The World Beneath Us
K The Invisible World
L Seas and Oceans
M Weather and its Work
N Astronomy
P Mathematics
Q Computers: An Introduction
R Working with Computers
S Through the Microscope
T How Everyday Things Work
U Great Discoveries and Inventions
V How Does it Work?
W Communications and Transport
X The World Today
Y Projects

internal fertilization **G** 28, 44–46, 47
language **W** 6
meaning **D** 6–7
plant-eating **F** 52–54
pre-adaptation **G** 58
skeleton **D** 10–11
species **G** 55
temperature **C** 60–61
water **F** 56
Mammoths **D** 41
Manchus **X** 36
Mandarin ducks **B** 26–27
Mandibles **B** 12; **E** 8
Manganese nodules **J** 53
Mangrove swamps **L** 23
Manta ray **C** 39
Mantis shrimps **C** 29
Mantle **L** 16
Manuscripts **W** 7
Maoris **X** 48, 49–50
Maples **A** 18, 35
Mapping, satellite **J** 50–52
Marble **J** 22
Marconi, Guglielmo **U** 52; **W** 9
Marcus, Siegfried **W** 21
Marijuana **Y** 42
Marine bristle worms **C** 15–16; **G** 27
Marine iguana **C** 55; **F** 11; **G** 59
Mariner probes **N** 25, 61
Marion's tortoise **C** 52
Marius, Simon **N** 53
Markhors **D** 52–53
Mars
axis **N** 25
diameter **N** 25
discover of **N** 10
distance from Sun **N** 25
Mariner probes **N** 25
moons **N** 25
revolving of **N** 16
Viking probes **N** 25
Marshall Islands **X** 51
Marsupial mice **D** 14
Marsupials **D** 14–15; **F** 50; **G** 60; **S** 7
Marten **D** 34
Martins **B** 51
Mass death, dinosaurs **F** 42–43
Mass production **U** 11
Massachusetts Institute of Technology **R** 18
Mastodons **D** 41
Mathematicians
ancient Egyptian **P** 26–27
early **P** 24

▼ Mammals come in many sizes. The white whale of the Arctic Ocean is among the largest.

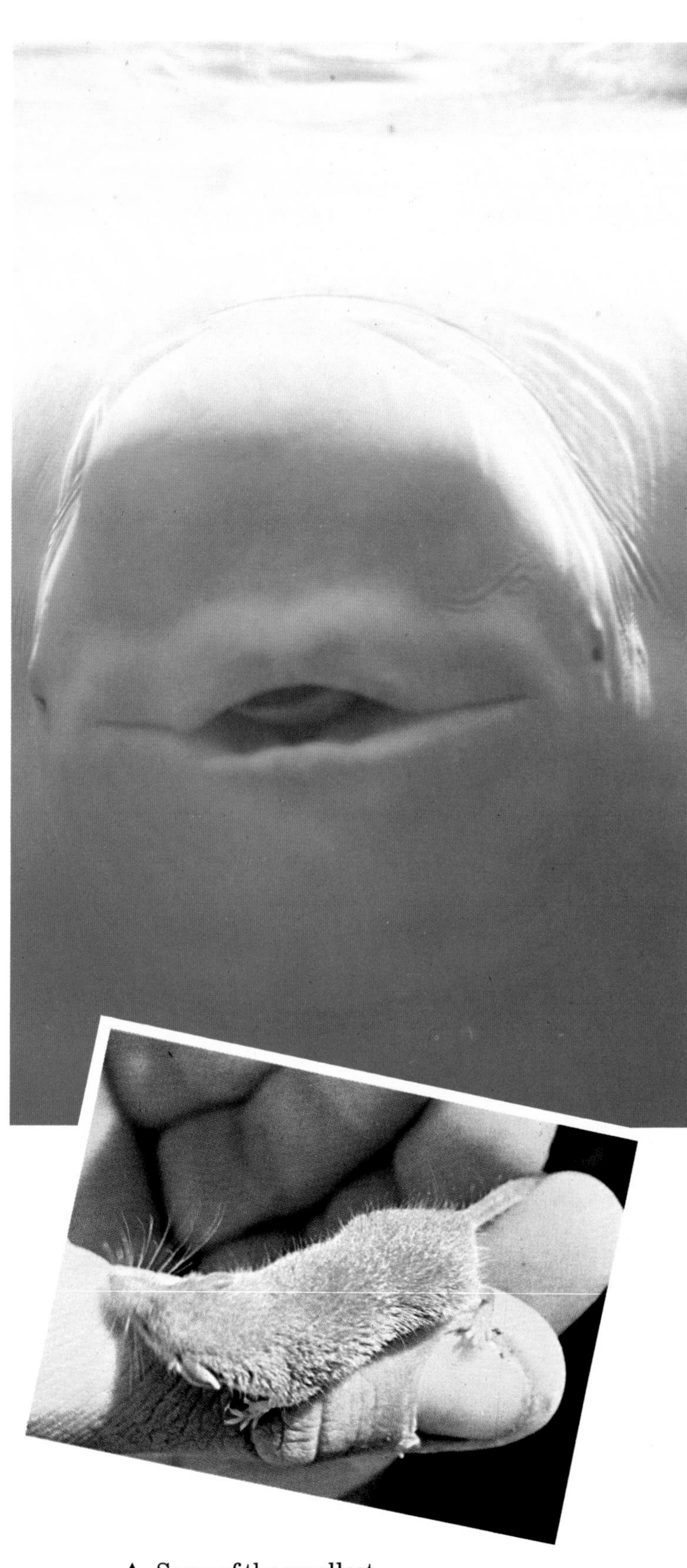

▲ Some of the smallest mammals are shrews. The human thumb on which this shrew is sitting shows just how tiny it is.

Greek **P** 28–29
Mathematics (volume **P**)
applied **P** 24
as language **P** 6–7
as magic **P** 30–31, 34
astronomy and **P** 24–25
increasing importance of **P** 60
optics **P** 41
revolutions in **P** 38–43
theory of **P** 24
Matter
into energy **K** 30
meaning **K** 6
structure **K** 6
Mauchly, John **Q** 22
Mauritania **X** 44
Mauritius **X** 46
Maxillae **E** 8
McAdam, John **W** 14
Meadowlark **B** 50
Measles **S** 48
Measuring
basic **P** 26
cubic **P** 16
cubit **P** 26
instruments **P** 44
modern systems **P** 44–45
need to measure **P** 26–27
scales **P** 26
time **P** 45
Mechanical arm **V** 60, 61
Medical detectives **S** 52–53
Medical diagnosis **Q** 47
Medical school **I** 14
Medical vocabulary **I** 7
Medicinal leech **C** 17
Medicine (volume **I**)
Egypt **I** 10
future of **I** 61
Greece **I** 11–12
in the past **I** 10–12
meaning **I** 8
progress **I** 13–14
Rome **I** 12
Mediterranean fruit fly **E** 16
Mediterranean sailors **W** 34
Mediterranean sea **L** 10
Meganeura **F** 22
Megapodes **B** 34
Meiosis **G** 50
Melanesia **X** 51
Melanesians **X** 40, 51
Melting **K** 14
Membrane, cell **A** 27; **G** 47; **H** 12, 14
Memory **Q** 13; **R** 9, 10–11
Mendel, Gregor **G** 48–49
Mendeleev, Dmitri **U** 12
Mental illness **I** 11, 38–39
Mercator, Gerardus **P** 35
Mercury
diameter **N** 18
discovery of **N** 10
metal **J** 19, **K** 18
orbit **N** 18
phases **N** 18
surface **N** 18
Mergansers **B** 27
Mesopotamia **W** 6; **Y** 8, 9
Mesosaurus **F** 24
Mesozoic era **F** 19, 26–28, 30
Messier, Charles **N** 43, 46
Mestizos **X** 14, 18, 22, 23
Metal
alloy **J** 32
atoms **S** 61
crystals **S** 34
heavy **K** 18
ores **J** 30–31, 32
Metalworking **U** 8
Metamorphosis
complete **E** 14, 15
frogs **C** 46
incomplete **E** 14, 15
insects **C** 34, **G** 29, 33
meaning **E** 14
Metcalf, John **W** 14
Meteorites **J** 8, 10; **N** 31
Meteorologists **M** 12, 30
Meteors **N** 31
Mexico **X** 13, 15, 18
Mice
dormice **D** 30
harvest **D** 26
in homes **S** 25
northern pygmy **D** 26
white **D** 26
Michaux, Ernest **T** 6
Michaux, Pierre **T** 6
Michelin brothers **W** 22
Micro-organisms **I** 16, 17
Microbes
description **S** 52–53
fighting **S** 50–51
future use **S** 37
germ theory of disease **I** 17
good **S** 46
harmful **S** 46–47
plagues **S** 48–49
Microcomputer **P** 56; **Y** 11
Micromata roseum spider **C** 36
Micronesia **X** 51
Micronesians **X** 42
Microphone
different types **V** 48
electrostatic **V** 48
modern **V** 49
ribbon **V** 47, 48
work of amplifiers **V** 48
Microprocessor **Q** 9, 13, 16; **R** 10
Microscopes **H** 8, 9; **I** 15–16, 31; **U** 14–15
Microscopic world volume **S**
Microwriter **R** 25
Mid Ocean Ridge **J** 29; **L** 16
Middle Ages **W** 7
Midnight sun **M** 51
Midwater shark **C** 38
Midwife toad **C** 47
Migrations **B** 28, 36; **Y** 29
Milk
mammal's **D** 7
pigeon's **B** 38
snake **C** 56
what is **K** 12
Milky Way **K** 6; **N** 12, 35, 47

THE WORLD'S LONGEST RIVERS

The table shows rivers that are over 4,000 km (2,500 mi) long. Many of them have hyphenated names because different stretches are called by different names. The length of a river gives no clue as to the amount of water it discharges; the Amazon is the greatest of rivers, with an outflow nearly 10 times that of the Mississippi-Missouri.

	km	mi	continent
Nile	6,695	4,160	Africa
Amazon	6,516	4,049	South America
Yangtze	6,380	3,964	Asia
Mississippi-Missouri	6,019	3,740	North America
Ob-Irtysh	5,570	3,461	Asia
Yenisey-Angara	5,550	3,449	Asia
Hwang Ho	5,464	3,395	Asia
Zaire	4,667	2,900	Africa
Parana	4,500	2,796	South America
Mekong	4,425	2,750	Asia
Amur	4,416	2,744	Asia
Lena-Kirenga	4,400	2,730	Africa
Mackenzie	4,250	2,641	North America
Niger	4,030	2,504	Africa

THE WORLD'S LARGEST LAKES AND INLAND SEAS

Inland bodies of water of area greater than 17,000 sq km (6560 sq mi) are shown here. Two of these, the Caspian Sea and the Aral Sea, are salty, because they are not drained by rivers – they lose their water by evaporation, so the salt is left behind.

	sq km	sq mi	continent
Caspian Sea	371,000	143,240	Asia
Lake Superior	83,270	32,150	North America
Lake Victoria	68,800	26,560	Africa
Aral Sea	65,500	25,290	Asia
Lake Huron	60,700	23,440	North America
Lake Michigan	58,020	22,400	North America
Lake Tanganyika	32,900	12,700	Africa
Great Bear Lake	31,790	12,270	North America
Lake Baykal	30,500	11,780	Asia
Great Slave Lake	28,440	10,980	North America
Lake Erie	25,680	9,915	North America
Lake Winnipeg	24,510	9,460	North America
Lake Malawi	22,490	8,680	Africa
Lake Ontario	19,230	7,425	North America
Lake Ladoga	18,390	7,100	Europe
Lake Balkhash	17,400	6,720	Asia

KEY TO TITLES OF VOLUMES

A Plants
B Birds
C Cold-blooded Animals
D Warm-blooded Animals
E Insects and their Relatives
F The World Before Man
G The Beginnings of Life
H The Human Body
I Disease and Medicine
J The World Beneath Us
K The Invisible World
L Seas and Oceans
M Weather and its Work
N Astronomy
P Mathematics
Q Computers: An Introduction
R Working with Computers
S Through the Microscope
T How Everyday Things Work
U Great Discoveries and Inventions
V How Does it Work?
W Communications and Transport
X The World Today
Y Projects

Miller, Stanley **F** 13
Millions, magnifying by **S** 60–61
Millipedes
- ancestors of **F** 22
- distinguished from centipedes **C** 26; **E** 58, 59
- early **F** 22
- female **E** 58–59
- meaning **C** 26; **E** 58
- ringed bodies **C** 34
- size **C** 26
- stink glands **E** 58

Mind power **K** 58–61
Mindanao Trench **L** 8
Minerals
- Antarctic **J** 53
- collecting **Y** 18–19
- collection from earth's surface **J** 39
- crystals **K** 17; **S** 34
- deep sea **J** 53
- deposits **J** 30
- dredging **J** 44
- first users **J** 30–31
- formation **J** 19, 20
- from moving crust **J** 33–34
- from seas and rivers **J** 44
- hidden beds **J** 36
- identified in rocks **J** 11
- importance of **J** 6
- searching today **J** 35–38
- tumble polishing **Y** 18
- where continents collide **J** 34
- where continents split **J** 33–34

Mining
- deep mines **J** 42–43
- drift mine **J** 42
- first mines **J** 31
- hazards of **J** 43
- open-cast **J** 40–41, 42
- sea **L** 59
- shaft mine **J** 42
- slope mine **J** 42

Mink **D** 34
Miocene **F** 46
Mirage **Y** 39
Mirrors
- concave **T** 40, 41
- construction **T** 38
- convex **T** 40–41
- curved **T** 38
- distorting **T** 38
- formation of images **T** 38, 40
- magnifying **T** 38
- rear-view **T** 38
- reflecting telescope **T** 41
- types **T** 38
- writing **Y** 9

Mistletoe **A** 6, 58; **G** 23
Mites **S** 24, 25
Mitosis **G** 49
Moas **F** 48
Mobile robots **R** 60
Möbius, August **P** 42
Mockingbirds **B** 53
Modem **R** 45, 46
Moh's Scale **Y** 19, 27
Molecules
- complicated **K** 37
- made up of **S** 59; **U** 12
- meaning **K** 12
- metal **K** 37
- of heredity **K** 43
- protein **K** 40, 42, 43
- simple **K** 37
- solids **K** 16, 20
- symmetry **P** 21
- through the microscope **S** 60–61

Moles **D** 16
Molluscs
- bivalve **C** 18–19, 24; **F** 16; **L** 37–38
- gastropod **C** 20–21; **F** 16, 17; **L** 37; **Y** 20
- major groups **F** 16–17
- parthenogenesis **G** 13

Molybdenum **Y** 32
Monaco **X** 30
Monarch butterfly **E** 15
Monasteries **I** 14
Mongoloids **X** 10, 11, 13, 35, 48, 55
Mongooses **D** 35
Monitor **R** 25
Monitor lizards **C** 54
Monitor socket **Q** 11
Monkeys **D** 20–21; **G** 23
Monocotyledons **A** 19, 28
Monotremes **F** 49–50; **G** 44–45
Monsoon
- climates **M** 40–41
- regions **M** 40
- summer **M** 40
- winter **M** 40

Mont Blanc **W** 16
Mont St Michel, France **J** 17
Montana, USA **J** 14
Montgolfier, Jacques **Y** 44
Montgolfier, Joseph **Y** 44
Montgolfier brothers **W** 46
Montserrat **X** 19
Moon
- Apollo mission landing sites **K** 8; **N** 21, 22
- atmosphere **N** 23
- buggy **U** 37
- diameter **N** 23
- distance from earth **N** 20
- earth–moon system **N** 20–23
- eclipses **N** 24
- effect on tides **L** 10
- exploration **V** 55
- features **N** 23
- gravitational pull **N** 23
- maria regions **N** 23
- phases **N** 20–21, 23
- recognizing **Y** 22, 23
- revolving of **N** 20
- rock **S** 34–36
- rock samples **N** 23
- rotation **N** 23

Moose **D** 50, 51
Morocco **X** 44
Morse, Samuel **W** 8
Morse code **W** 8, **Y** 16
Mosasaurs **F** 38
Mosquito **E** 10, 13, 17; **I** 34; **S** 8
Mosses **A** 6, 12; **G** 16, 17
Moths **E** 9, 15, 34–36; **G** 11, 19, 33; **S** 8
Motion picture
- animated **T** 36
- camera **T** 36–37
- shooting **T** 36
- sound track **T** 37
- stunt scenes **T** 36

Motmots **B** 45
Motor cars *see* Cars
Motorcycle
- first **V** 16
- gears **V** 17
- piston engines **V** 16–17
- racing **V** 16
- shock absorbers **V** 17
- sidecar **V** 17

Mouflon **D** 53
Mould **S** 25
Moulting **E** 14, 15
Mound-builders **B** 34
Mount Etna, Italy **J** 25
Mount Everest **L** 8
Mount Vesuvius, Italy **J** 26
Mountain
- climates **M** 52–53
- cloud and rain **M** 53
- life **M** 53
- winds **M** 52–53

Mouth-brooding **G** 36
Movement **H** 44–45
Mozambique **X** 46
Mudpuppies **C** 46
Mudskippers **C** 40; **F** 2–3; **G** 37
Mulattos **X** 14, 22
Müller, Erwin **U** 15
Multiple sclerosis **I** 60
Multiplication **P** 12
Muntjac **D** 50
Muscles
- birds **B** 9
- cells **H** 31
- different kinds **H** 30–31
- function **H** 30–31
- heart (cardiac) **H** 31
- involuntary **H** 30, 31
- map **H** 11, 30
- movement **H** 44–45
- skeletal **H** 30, 31
- voluntary **H** 30, 31

Museums **J** 56
Mushrooms **A** 10–11; **S** 31
Music
- home micro **R** 38
- making **R** 36–38
- processing **R** 37
- sampling **R** 37
- sequencers **R** 36, 37
- studio sounds **R** 36–37
- synthesizers **R** 36–37

Musk-ox **D** 53, 57
Muslims **X** 37, 38, 39, 44
Mussels **C** 18, 31; **G** 26, 27
Mutations **G** 50, 57, 59
My Talking Computer **R** 40
Mycelium **A** 10
Myelin sheath **H** 32
Myriapods **C** 27

N

▲ Navigation is now a very exact science, frequently involving the use of satellites. The very earliest sailors had little to guide them except the sun, the moon and the stars.

Nail-sitter **K** 59
Nails **H** 46
Names **G** 56
Namibia **X** 40, 44
Napier, John **P** 12
Natural fertilizers **A** 42
Natural selection **G** 56–57, 58
Nature **S** 6–7
Nautiloids **F** 16
Navigation **P** 22, 35–37; **W** 42–43, 58–59
Navigation aids **L** 53
Navigation instruments **U** 10
Neanderthal man **X** 7–8
Nebulae **N** 44–46
Needles **A** 15
Negatively-charged particles **Y** 30
Negritos **X** 10, 48, 51
Negroid **X** 10, 11, 40
Nelson, Horatio **W** 38
Nepal **X** 38
Neptune
- diameter **N** 28
- discovery of **N** 10
- moons **N** 28
- orbit **N** 11
- rotation **N** 28

Nereid **N** 28
Nerve cells **H** 14; **S** 38, 39
Nerve system **E** 12
Nervous system **H** 11,
Nest-building **B** 13, 34; **E** 43, 44, 45, 46, 47, 58; **G** 36, 41, 43
Nest-parasites **B** 40
Netherlands **X** 19, 22, 28, 49
Nettle animals **C** 12–13, 21
Neurones **H** 32–33, 34
Neuroses **I** 38
New Caledonia **X** 51
New World vultures **B** 29
New Zealand **A** 13; **X** 48, 49, 51
Newman, Max **Q** 22
Newton, Sir Isaac **N** 16, 23, 32, 40, 52, 55; **P** 40–41; **U** 38, 48, 51
Newton's disk **Y** 36
Newton's laws **N** 17
Newts **C** 44–45; **G** 37
Niagara Falls **U** 44
Nicaragua **X** 18
Nickel **Y** 32
Nidicolous birds **G** 43
Nidifugous birds **G** 43
Niger **X** 44
Nigeria **X** 45
Nighthawk **B** 42; **C** 58
Nightingale **B** 13, 54
Nightingale, Florence **I** 50
Nightjars **B** 42; **C** 58
Nilo-Hamites **X** 40
Nitrogen **K** 10, 11
Nitroglycerine **V** 42
Nobel, Alfred **V** 42; **W** 33
Nobel, Immanuel **V** 42
Noctiluca **S** 32
Non-crystals **K** 16
Nordic people **X** 27
Norris, William **W** 27
North America **A** 15, 36; **X** 10, 12, 13–14, 27, 55
North Korea **X** 36
North Mariana Islands **X** 51
North-eastern Asia **X** 36–37
Northern Africa **X** 10, 44–45
Northern Europe **X** 28–30
Northern Ireland **X** 29
Northern Lights **Y** 37
Northern pygmy mice **D** 26
Norway **X** 27, 28
Nose **H** 42
Notebook, ecology **Y** 60–61
Nothosaurs **F** 38
Nuclear
- energy **U** 54; **V** 36
- explosives **V** 42
- fission **U** 54; **V** 37
- fusion **U** 54
- power station **U** 55; **V** 36, 38
- reactor **U** 54; **V** 36–39, 50
- submarine **L** 14, 15

Nucleic acid **I** 30, 31; **K** 43
Numbers
- groups **P** 10
- imaginary **P** 11
- irrational **P** 11
- negative **P** 10
- rational **P** 10–11
- systems **P** 10, 11
- writing of **P** 10

Numerals **Y** 9
Nurses **I** 50–51
Nursing **I** 50–51
Nutcrackers **V** 33, 35
Nutrients **I** 40–41
Nuts **G** 23
Nyala **D** 59

O

▲ Orchids grow in most tropical and temperate parts of the world. There are so many species that they are highly prized by plant collectors. Sadly, many wild orchids are becoming rare.

Oak trees **A** 18; **G** 23
Oasis **J** 48
Oats **U** 18, 19
Observatories **N** 56, 60; **Y** 23
Ocean (volume **L**)
abyss **L** 16
area covered by **L** 6, 7–8
currents *see* Currents
depth **L** 8
floor **J** 10; **L** 15–16
highways **L** 53
pressure **L** 8
relative sizes **L** 7
salty **L** 8
sunfish **G** 37
temperature of water **L** 8
Oceania **X** 26, 27, 48
Oceans **S** 32–33
Octopus **C** 12, 18, 22
Ohain, Dr Hans von **W** 53
Oil **J** 46
Okapi **D** 55
Old World vultures **B** 29
Olgas, Australia **J** 15
Oligocene **F** 46
Oligokyphus **F** 25
Oman **X** 37
On/off switch **Q** 17
Onion **G** 25
Onion fly **E** 38
Ooze
calcareous **L** 15
diatom **L** 15
radiolarian **L** 15
Open-cast mining **J** 40–41, 42
Operating rooms **I** 52–53
Operating system **Q** 13, 16
Operations **I** 8, 48, 56–57
Ophiderpeton **F** 23
Opossums **D** 7, 14, 15; **S** 7
Optic nerve **H** 35
Optical illusions **Y** 38–39
Opticon **R** 51
Orangutan **D** 21
Orbit **V** 52
Orchids **A** 5, 18, 19, 22, 32, 51; **G** 22, 61
Ordovician period **F** 14–15, 16
Organs **H** 11; **S** 42–43
Orioles **B** 56
Ornamental trees **A** 51
Ornithischians **F** 30, 35
Ornithomimus **F** 31
Ornithopods **F** 34–35
Oscillatoria **S** 20
Osmium **K** 18
Osmosis **A** 26, 27
Osprey **B** 32
Osteodontornis **F** 48
Ostrich **B** 16
Otters **D** 34, 35
Otto, Nikolaus August **V** 31; **W** 20
Output unit **Q** 9
Ova **G** 11, 53
Ovenbirds **B** 47
Overbalancing wheel **V** 45
Overeating **I** 60
Owls **B** 10, 11, 15, 41
Ox-eye daisies **A** 7
Oxidation **K** 34, 36
Oxides **J** 20
Oxpecker **B** 12
Oxygen **A** 24, 25; **C** 10; **E** 10; **F** 13; **K** 10–11, 34, 36
Oyster **C** 18
Ozone **F** 13; **K** 11

P

▲ Paraguay: the presidential palace in the capital city, Asuncion.

Pachydyptes **F** 48
Pacific Ocean **L** 7–8, 10, 14, 16, 24
Pain
  threshold **I** 24
  weapons against **I** 24–27
Pairing **B** 13, 21
Pakistan **X** 9, 38
Palaeocene **F** 46
Palaeozoic Era **F** 13, 14, 25
Pallas **N** 30
Palms **A** 19, 21, 22
Palolo worm **G** 27
Panama **X** 18, 20
Panama Canal **L** 54
Pancreas **H** 29
Pandas **D** 32
Pangaea **F** 25, 27; **J** 34
Pangolins **D** 23
Panhard, Rene **W** 21
Pantograph **Y** 8
Paper
  nautilus **C** 22
  production **U** 22–23
  pulp **U** 23
Papert, Seymour **Q** 48, 49
Papua New Guinea **X** 48, 51
Parachute **V** 21
Paraguay **X** 22
Parallel evolution **G** 60
Parallel interface **Q** 10
Paramecium **C** 9; **S** 15
Parasites **A** 6, 58–59; **B** 40; **I** 32–34; **S** 46
Parasitic worms **I** 33–34
Parathyroid gland **H** 28
Parent **G** 6
Parkes, John **U** 30
Parks **Y** 61
Parrakeets **B** 39
Parrotlets **B** 39
Parrots **B** 6, 11, 39; **G** 23
Parthenogenesis **G** 13
Particles **K** 20, 28; **S** 38
PASCAL **Q** 48
Pascal, Blaise **P** 38–39, 40; **Q** 20, 22; **U** 10
Pasha butterfly **C** 7, 34
Pasteur, Louis **I** 13, 17, 18, 22, 23; **S** 50, 53
Pateurization **I** 17; **S** 50
Patents **V** 45
Peacock butterfly **E** 34
Peacock worms **C** 15, 16
Peanuts **A** 42
Pearly nautilus **C** 22
Peas **A** 25, 42; **G** 48–49, 51
Peat **K** 37
Peccary **D** 47
Pediastrum **S** 20
Pelicans **B** 23
Pelycosaurs **F** 25
Penguins **B** 8, 13, 18–19
Penicillin **I** 24, 25, 32
Penicillium mould **S** 26
Penman plotter **R** 34
Pennines, England **J** 11
Pens, light **Q** 59; **R** 34
People Mover **R** 60
Peppered moth **G** 58–59
Peregrine **B** 32
Peripherals **Q** 6
Periscope **Y** 34
Periwinkles **C** 21
Permian period **F** 15, 25; **J** 45
Perpetual motion **V** 44–45
Peru **X** 20, 22
Pests **E** 16–17
Pet keeping **Y** 56–59
Petals **A** 28, 29; **G** 18, 19
Petrels **B** 22
Pheasant **B** 14, 33
Philippines **X** 39
Philodina **S** 22
Philosophers **P** 10
Phobos **N** 25
Phonemes **R** 41
Phorusrhacos **F** 47, 48
Photomicrograph **S** 26, 31, 34–36
Photonic wand **R** 50, 51
Photosynthesis **A** 6, 24–25, 26, 27; **F** 13; **G** 14; **Y** 48, 49
Physical tests **Y** 26–27
Physicists **I** 54; **P** 7
Physics
  mathematics applied to **P** 41
  science puzzles **Y** 6–7
  tests **Y** 26–27
Physiology **H** 10–11; **I** 14
Physiotherapists **I** 54–55

KEY TO TITLES OF VOLUMES

A Plants
B Birds
C Cold-blooded Animals
D Warm-blooded Animals
E Insects and their Relatives
F The World Before Man
G The Beginnings of Life
H The Human Body
I Disease and Medicine
J The World Beneath Us
K The Invisible World
L Seas and Oceans
M Weather and its Work
N Astronomy
P Mathematics
Q Computers: An Introduction
R Working with Computers
S Through the Microscope
T How Everyday Things Work
U Great Discoveries and Inventions
V How Does it Work?
W Communications and Transport
X The World Today
Y Projects

Phytoplankton **A** 6; **L** 28; **Y** 55
Phytosaurs **F** 28
Piaget, Jean **Q** 49
Piazzi, Giuseppe **N** 30
Picture language **W** 6–7
Pigeons **B** 11, 38
Pigs **D** 47
Pikas **D** 24
Pindus mountains, Greece **J** 50
Pine hawk moth **E** 36
Pine needle **S** 29
Pines **A** 14, 15, 16
Pintail **B** 7
Pioneer probes **N** 60
Pipistrelle bat **D** 19
Pitcher plant **A** 54, 55
Pituitary gland **H** 28
Placcoderms **F** 19
Placentals **G** 46
Placodonts **F** 38
Plague **E** 16; **S** 48
Planarians **S** 10
Planets
  brightness **N** 40
  discovery of **N** 10
  giant **N** 26–29
  inner **N** 18–19
  minor **N** 30
  names **N** 10
  orbits **N** 10, 11, 16–17
  recognizing **Y** 22
  sizes **N** 11
Plankton
  as shark food **C** 38
  chlorophyll **L** 28
  fishing **Y** 54–55
  food chains **L** 28–29
  food pyramids **L** 28
  in oceans **S** 32
  phosphorescence **L** 28
  phytoplankton **A** 6; **L** 28; **Y** 55
  plants and creatures **L** 29
  zooplankton **L** 28
Plants (volume **A**)
  annual **G** 11
  cells **A** 6
  early **F** 20–21, 46
  fibres **A** 6
  flowering **G** 18–21
  growing **Y** 48–49
  lice **E** 30–31
  propagation **Y** 50–51
  reproduction **G** 14–25
  skeleton **S** 30, 31
  species **G** 55
  swimming **S** 15
  tetraploidy **G** 58
  water **S** 10
  work of **J** 24
Plasma **K** 20
Plastics
  early **U** 30
  modern **U** 31
  polyvinyl chloride **U** 31
  synthetic **U** 30–31
Plated lizard **C** 55
Platelets **H** 21
Platinum **K** 18
Platypus **D** 12–13; **G** 44, 46
Pleistocene **F** 46, 61
Plesiosaurs **F** 38
Pliers **V** 33
Pliocene **F** 46
Plotopterids **F** 48
Plotters **Q** 57; **R** 35
Plumatella **S** 22
Pluto **N** 10, 11, 29
Plutonium **V** 38
Pneumonia **I** 28–29
Poisons **A** 48–49; **G** 57
Poland **X** 31–32
Polar bears **L** 47
Polar climate
  life in **M** 50
  permafrost **M** 50
  summer **M** 50
  winter **M** 50
Polar easterlies **M** 21, 22
Polar front **M** 15
Polecat **D** 34
Polio **I** 18
Pollen **A** 28–30
Pollen grains **S** 28
Pollination
  artificial **S** 29
  by insects **E** 18; **G** 19
  cross **A** 33
  plants **G** 19, 21
  self **A** 33
  stages **A** 32
  wind **A** 32–33
Pollution **A** 56; **L** 61; **M** 33; **X** 59
Poltergeists **K** 57, 58
Polygons **P** 16, 18
Polynesia **X** 51
Polynesians **X** 10, 15, 48, 49
Polyp **C** 12, 13
Polythene **U** 30
Pond building **Y** 52–53
Pond skater **C** 33
Ponds **S** 16, 18, 22
Pong **R** 18, 19, 21
Poor-will **B** 42; **C** 58
Poppies **A** 7
Porcelain **U** 20–21
Porcupine **D** 9, 26, 27
Porpoises **D** 60
Portugal **X** 22, 30, 36, 40
Potatoes **A** 40–41; **G** 24–25
Pouched mammals *see* Marsupials
Poultry **B** 11
Power
  connector **Q** 11
  human beings **K** 50
  socket **Q** 17
  supply **Q** 15
Prairie chicken **B** 33
Prawns **C** 30; **E** 52, 53; **L** 34
Pre-adaptation **G** 58
Precambrian **F** 13
Preening **B** 7
Prehistoric insects **E** 20
Prevention **I** 61
Priestley, Joseph **U** 34
Primates **D** 20–21; **X** 6
Primroses **A** 33; **G** 18
Principe **X** 45
Printed circuit boards **Q** 16
Printer
  daisy wheel **R** 27–27
  dot matrix **Q** 56–57; **R** 27, 35
  types **Q** 6, 56–57
  use **Q** 6, 13
Printing
  composing **U** 24
  invention **U** 10, 24; **W** 7
  mathematics **P** 35
  presses **U** 24–25
  processes **U** 25
  telegraph **W** 8
  works **U** 25
Privet hawk moth **E** 34
Probability
  empirical **P** 50
  theory **P** 40, 50
Process **Q** 35
Processor unit **Q** 9
Program
  counter **Q** 12, 13
  examples of BASIC **R** 14–15
  loading **R** 9
  user friendliness **Q** 34–35; **R** 16
Programmers **Q** 36–37
Programming
  language **Q** 28–29
  meaning **Q** 7
  ready-made programs **Q** 38–39
  robots *see* Robots
  user friendly programs **Q** 34–35; **R** 16
  writing a program **Q** 30–34
Prolegs **E** 15
Pronghorn antelope **D** 58; **X** 57
Propagation of plants **Y** 50
Prospectors **J** 35, 36
Proteins **A** 24; **H** 51; **I** 30, 31, 40–41; **K** 40, 42, 43
Proterosuchians **F** 28
Protozans **G** 6, 8, 11
Protozoa **C** 8–9; **S** 14, 15, 22
Pseudopodia **S** 16, 17
Pseudosuchians **F** 28
Psychiatrists **I** 39
Psychokinesis **K** 58–59
Psychologists **I** 39
Psychosomatic disease **I** 39
Psychoses **I** 38–39
Psychotherapists **I** 55
Ptarmigan **B** 33
Pteranodon **F** 41
Pteraspids **F** 19
Pterodactylids **F** 41
Pterosaurs **F** 27, 28, 40–41
Ptolemy, Claudius **N** 8, 9, 16, 17
Pudu **D** 50
Puerperal fever **I** 22
Puerto Rico **X** 19
Puffballs **A** 10, 11
Puffer fish **C** 40
Pulley **V** 34, 35
Pulsars **N** 42–43
Puma **D** 36
Pumps
  bicycle **T** 14, 15
  centrifugal **T** 15
  force **T** 14–15
  gear **T** 15
  jet **T** 15
  lifting **T** 14
  reciprocating **T** 15
  vane **T** 15
  windmills **T** 14
Punched cards **Q** 19
Pupae **C** 34; **E** 34; **G** 33
Pure mathematics **P** 35
Purse sponge **C** 10, 11
Puss moth **E** 36
Puzzles, science **Y** 6–7
Pygmies **X** 8, 10, 40, 44
Pyramids **N** 6, 7
Pyromorphite **K** 17
Pythagoras **P** 28–29
Pythagoreans **P** 28–29, 31
Pythons **C** 57; **G** 39

KEY TO TITLES OF VOLUMES

A Plants
B Birds
C Cold-blooded Animals
D Warm-blooded Animals
E Insects and their Relatives
F The World Before Man
G The Beginnings of Life
H The Human Body
I Disease and Medicine
J The World Beneath Us
K The Invisible World
L Seas and Oceans
M Weather and its Work
N Astronomy
P Mathematics
Q Computers: An Introduction
R Working with Computers
S Through the Microscope
T How Everyday Things Work
U Great Discoveries and Inventions
V How Does it Work?
W Communications and Transport
X The World Today
Y Projects

Qatar **X** 38
Quadricycle **T** 6
Quadrilaterals **P** 16, 18
Quarantine **I** 14
Quartz grains **J** 21
Quasars **N** 49
Quaternary period **F** 46
Queen Elizabeth 2 **L** 52
Quetzalcoatlus **F** 41
Quills **B** 6; **D** 8, 9

▲ The *Queen Elizabeth 2* is used mostly as a cruise ship. Ships cannot compete with airplanes as a means of long-distance passenger transport.

R

▲ Railways face stiff competition from roads as a means of long-distance freight transport.

Rabbit
  cottontail **D** 24
  European **D** 24, 25
  marsh **D** 25
  Mexican pygmy **D** 25
  swamp **D** 25
Raccoons **D** 33
Racial discrimination **X** 46
Radar
  development **U** 52
  radio echoes **U** 52
  scanner **U** 53; **W** 43
  screen **U** 53

KEY TO TITLES OF VOLUMES

A Plants
B Birds
C Cold-blooded Animals
D Warm-blooded Animals
E Insects and their Relatives
F The World Before Man
G The Beginnings of Life
H The Human Body
I Disease and Medicine
J The World Beneath Us
K The Invisible World
L Seas and Oceans
M Weather and its Work
N Astronomy
P Mathematics
Q Computers: An Introduction
R Working with Computers
S Through the Microscope
T How Everyday Things Work
U Great Discoveries and Inventions
V How Does it Work?
W Communications and Transport
X The World Today
Y Projects

  set **U** 52
  uses **U** 52–53
Radiation **I** 47; **K** 30
Radio
  aircraft **T** 44
  amateur enthusiasts **T** 43
  amplifying the current **T** 44
  amplitude modulation **T** 43–44
  communications **W** 8–9
  construction **Y** 32
  disc jockey **T** 44
  frequency modulation **T** 44
  phones **V** 26
  telescopes **N** 56–59
  transistor **T** 54
  use **Y** 30
  walkman **T** 42, 49
  waves **T** 42–43
Radioactivity **K** 23; **V** 46–47
Radiotherapy **I** 55
Ragworms **C** 15
Rails and horses **W** 26
Railways **W** 26–33
Rain
  acid **M** 33
  amounts **M** 14
  convectional **M** 14
  drizzle **M** 14
  forests **A** 21–23, 60; **M** 38–39
  formation **L** 6
  frontal **M** 15
  heavy **M** 14
  light **M** 14
  moderate **M** 14
  monsoon **M** 40–41
  mountain **M** 53
  reasons for falling **M** 14
  relief **M** 15
  tropical rainy climates **M** 38–39
  types **M** 14–15
RAM see Random access memory
Ramapithecus **X** 6
Random access memory (RAM) **P** 56, 57; **Q** 10, 13, 16, 54; **R** 10, 11, 42
Rapid eye movement (REM) **K** 46, 48
Ratfish **L** 41
Rats
  black **D** 26
  desert **D** 5
  white **D** 26
Rattlesnakes **C** 57
Rayon **U** 29
Rays
  cosmic **K** 22
  heat **K** 22–23
  light **K** 22–23
  radio **K** 24
  vertebrates **C** 38, 39; **L** 40, 41
  X **K** 24
Re *see* Sun
Reaction time **H** 33
Read and write memory *see* Random access memory
Read only memory (ROM) **P** 56, 57; **Q** 11, 13, 17; **R** 10–11, 42
Reading **W** 7
Record
  compact disc **T** 47
  digital recording **T** 46–47
  making **T** 45
  player **T** 45, 46

stereo recording **T** 45
Red clay **L** 15
Red Cross **I** 7
Red Sea **J** 33; **L** 8
Redwoods **A** 16, 26
Reef builders **F** 17
Reflexes **H** 32
Refrigeration **T** 26; **U** 11
Refrigerator
construction **T** 26
domestic **T** 26
large scale **T** 26
types **T** 26
Registers **Q** 13
Reindeer **D** 50, 51
Relative density **Y** 27
Relativity, theory of **P** 42–43
Religion **X** 18, 22, 27, 32, 33, 35, 38, 39, 40, 44
REM *see* Rapid eye movement
Remote Maintenance Vehicle **V** 51
Reproduction
animal **G** 26
asexual **G** 6, 8, 11, 13, 24
chromosomes **G** 49, 50, 58; **H** 54, 55; **I** 35; **K** 43; **S** 44–45
meaning **G** 6
microscopic creatures **S** 15
plants **G** 14–25
sexual **G** 6, 11
vegetative **G** 24–25
Reproductive sytem, human **H** 11, 52–53, 54
Reptiles
alligators **C** 50
and the sea **L** 50–51
crocodiles **C** 49–50; **F** 29, 38
lizards **C** 6, 7, 54–55; **F** 31
mammal-like **F** 25
modern, in the making **F** 29
pre-adaptation **G** 58
reproduction **G** 38–39
rise of **F** 24–25
snakes **C** 56–57
socket toothed **F** 28
stem **F** 24–25
tortoises **C** 51–52; **F** 29
tuatara **C** 48
turtles **C** 51–53, 62; **F** 38
unusual **F** 28–29
Republic of Belau **X** 51
Republic of Ireland **X** 28, 29
Research **I** 14, 15, 17, 54
Reserved modulator **Q** 11
Respiratory system
birds **B** 9
human **H** 11, 12, 48–49
insect **E** 10–11
RF modulator **Q** 11
Rhamphorhynchids **F** 41
Rhea **B** 17
Rhinoceroses **D** 8, 44–45; **G** 44
Rhipidistians **F** 19
Rhizomes **G** 24
Rhododendrons **A** 32
Rhynchosaurs **F** 28–29
Rice **A** 37–38; **U** 18, 19
Riflewren **B** 52
Ringworm **A** 11; **S** 47
Rivaz, Isaac de **W** 20
River horse **D** 46
Rivers
bore **L** 11
deltas **L** 22; **M** 56
estuaries **L** 22
lowland **M** 56
salt pans **L** 22
spurs **M** 56
transport **W** 44
tributaries **M** 56
upland **M** 56
Roadbuilding **W** 14–17
Roadrunner **B** 40
Roads **W** 14, 16
Robber ants **E** 46
Robber crab **C** 31
Robert, Nicholas **U** 23
Robins **B** 13, 54
Robots
at work **V** 56, 60–61
arms **R** 54, 58–9
construction **Q** 52–53; **R** 58
controlling **Q** 50
gripping device **R** 58
handling device **R** 58
holding parts **R** 59
industrial **Q** 50–53
on the move **R** 60
pick and place **R** 58
programming **Q** 50
servant **Y** 14
stepper motors **Q** 50
uses **R** 54
Rocket **V** 29
Rockets
fuel **V** 18–19
how they work **U** 50–51
instruments **U** 50
launch **V** 52
multi-stage **U** 50
principles behind **V** 18
retro **V** 52
Saturn **V** 19–20
space **U** 50, 51
story of **U** 51
Thor Delta booster **U** 5
Rocks
clastic **J** 22
clues in **J** 18
collecting **Y** 20, 21
crystals **J** 19
deposition **J** 21
diagenesis **J** 21
different types **J** 19–22
dissolving **M** 55
erosion **J** 19
faults and folds in **J** 20
formation **J** 8
freeze-thaw effects **M** 55
grain size **J** 22
igneous **J** 19, 20
limestone **M** 55
metamorphic **J** 19, 22
mineral ingredients **M** 55
minerals **J** 11
porous **J** 18
predicting earth-quakes **J** 54–55
radiometric dating **F** 9
reproduction **G** 6
sedimentary **J** 19, 20–22
swelling and shrinking **M** 55
transportation **J** 20
weathering **J** 20; **M** 55
without fissures **J** 18
Rodents
as pests **D** 26
carriers of disease **D** 26
development **F** 54
groups **D** 26
teeth **D** 11, 27
Roe, Alliot V **W** 50
Roentgen, Wilhelm Konrad **U** 46
Roland GR600 **R** 37
Rollers **B** 45
ROM *see* Read only memory
Roman
alphabet **W** 7
roads **W** 14
transport **W** 19
Romania **X** 31
Romans **Y** 27, 32
Rome **X** 27, 31
Rooks **B** 11, 61
Roots **A** 26–27, 40
Rope trick **K** 58
Rorschach, Hermann **I** 38
Rose's alloy **Y** 33
Rosellas **B** 39
Roses **A** 18, 51
Rotifers **S** 22, 23
Rubber
discovering **U** 34
man-made **U** 34–35
natural **U** 34
tree **U** 34, 35
Ruby **K** 16
Rudist **F** 17
Rudolff, Christoff **P** 35
Runners **G** 24
Running programs **R** 15
Russia *see* Union of Soviet Socialist Republics
Rust **K** 34
Rutherford, Ernest **U** 12, 13; **V** 46
Rwanda **X** 46

STANDING ROOM ONLY

Mankind's increasing mastery of the environment has been reflected in the size of the world's population, which has been rising since prehistoric times. There has been a spectacular increase since the Industrial Revolution of the late 18th century. This table shows how the time needed for the world's population to double has shrunk, until it is now about 30 years.

year	population (millions)
BC 300	125
AD 1	175
900	250
1250	350
1625	500
1750	700
1830	1000
1880	1400
1930	2000
1960	2800
1980	4000

S

SI system **P** 45
Safari ants **E** 47
Sahel **X** 44–45, 57
Sailing ships **W** 34–38
Sailplanes **V** 14
Salamanders **C** 44–45, 46; **G** 13, 37, 38
Sales of computers
 boom **R** 6–7
 crash **R** 7
 problems **R** 7
Saliva **H** 24
Sally lightfoot crab **C** 32
Salmon **C** 40, 43
Salt **K** 36, 37
Salvarsan **I** 24
Salyut stations **V** 56
Sampling **R** 37
San Marino **X** 31
Sand **J** 21, 22
Sand dollars **L** 35, 37
Sandpiper **B** 36
Sandstone **J** 19
Sandstorms **M** 45
São Tomé **X** 45
Saprophytes **A** 6, 10
Sarcomas **I** 46
Sargasso Sea **A** 9; **L** 12
Sargasso weed **L** 26
Satellites
 communications **V** 53–54; **W** 12, 13
 computer uses **Q** 45
 mapping **J** 50–52
 oceanography **L** 54
 unmanned **V** 52–53
 uses **V** 53
Saturn
 atmosphere **N** 27
 diameter **N** 27
 discovery of **N** 10
 moons **K** 28; **N** 27
 orbit **N** 11
 probes **V** 56
 rings **N** 27; **V** 57
 rotation **N** 27
Saturn V rocket **V** 19–20
Saudi Arabia **X** 37
Saurischians **F** 30, 35
Saurolophus **F** 35
Sauropods **F** 32–33
Scales
 butterflies' **E** 34
 weighing **P** 45
Scallop **C** 18, 19
Scandinavia **X** 28
Scarab beetle **E** 50, 51

◀ A spider in its web. Spiders have eight legs so they are not classified as insects.

Scavengers **E** 18
Schizophrenia **I** 39
Schlawlow, Arthur **U** 58
Schmidt, Dr Wilhelm **W** 28
Schönbein, Christian **U** 29
Schwabe, Heinrich **N** 14
Science puzzles **Y** 6–7
Scientific discovery **U** 10
Scientific visions **K** 56–57
Scorpion
 courtship dance **E** 60–61
 early **F** 22
 food **E** 60
 pincers **E** 60
 sting **E** 60
Screen grid **Q** 14
Screen phosphor **Q** 14
Screws
 screw-jack **T** 18
 thread **T** 18
 types **T** 18
 woodscrew **T** 18
 working **V** 35
Scrub-bird **B** 49
Scuba diver **K** 19
Sea (volume **L**)
 anemones **C** 12, 28; **G** 8; **L** 30, 31, 33; **S** 12, 13
 birds **L** 50
 changes at **J** 24
 cleaning **L** 61
 cows **F** 56
 cucumbers **C** 25; **L** 36, 37
 exploration **L** 54
 ice *see* Sea ice
 lettuce **C** 21
 life below **F** 14–15
 lilies **C** 15
 pressure **L** 8
 salt **L** 8
 scorpion **F** 14, 15
 shades of **L** 8
 slugs **C** 21; **L** 37–39
 snails **C** 21, 22; **F** 16; **L** 37
 snake **C** 56
 squirts **L** 30–31
 temperature of water **L** 8
 urchins **C** 24; **F** 9, 15; **G** 26–27; **L** 35, 37
 water **L** 6
 worms *see* Worms
Sea ice
 castellated bergs **L** 14
 coping with **L** 15
 floes **L** 13, 14
 formation **L** 13
 frazil **L** 13
 icebergs **L** 14–15
 icebreaker **L** 13, 15
 pack ice **L** 13
 tabular bergs **L** 15
Sea-fans **C** 12, 13
Sea-firs **C** 12, 13
Sea-hare **C** 21
Sea-hedgehog **C** 24
Sea-horse **C** 40, 42; **G** 36
Seagulls **B** 31, 37
Sealions **D** 38
Seals **D** 38–39; **F** 56; **L** 46–47
Seasons
 causes **M** 36–37
 division of year into **M** 36
 sunlight **M** 36, 37
Seaweed
 as food **C** 21; **L** 59
 brown **A** 9; **G** 14
 different kinds **L** 26–27
 green **A** 8
 holdfasts **L** 26, 27
 limp **L** 11
 red **A** 9
 survival **L** 27–28
Secretary bird **B** 31
Sedges **A** 19
Sediments **J** 22
Seeds **A** 34–35; **G** 22–23
Seeing
 method of **H** 36–37
 optic nerve **H** 35
Seismic prospecting **J** 35, 36–37
Seismic waves **J** 9
Seismographs **J** 9
Selenium **K** 16
Self-pollination **G** 19
Semaphore **Y** 16, 17
Semaphore telegraph **W** 8
Semiconductors **T** 54; **U** 56
Semmelweis, Ignaz **I** 22, 23
Senegal **X** 45
Sense-organs
 birds **B** 10–11
 insects **E** 12–13
Senses **Y** 40–43
Sensors **R** 60
Sentence Maker **R** 41
Sepals **A** 28; **G** 18, 19
Septic tank **T** 23
Sequence control

KEY TO TITLES OF VOLUMES

A Plants
B Birds
C Cold-blooded Animals
D Warm-blooded Animals
E Insects and their Relatives
F The World Before Man
G The Beginnings of Life
H The Human Body
I Disease and Medicine
J The World Beneath Us
K The Invisible World
L Seas and Oceans
M Weather and its Work
N Astronomy
P Mathematics
Q Computers: An Introduction
R Working with Computers
S Through the Microscope
T How Everyday Things Work
U Great Discoveries and Inventions
V How Does it Work?
W Communications and Transport
X The World Today
Y Projects

 register **Q** 13
Sequencers **R** 36, 37
Sequoia National Park, California **A** 16
Sequoias **A** 16
Serial port **Q** 17
Serpulid worm **C** 16
Servant robots **Y** 14
Severn bore **L** 11
Sewage **L** 61
Sewage works **T** 22–23
Sex cells **S** 44
Sex changes **G** 13
Sex glands **H** 29, 52
Sex instinct **K** 55
Sextants **P** 37
Sexton beetles **E** 50
Sexual reproduction **G** 6, 11; **H** 11, 52–53, 54; *see also* Fertilization
Seychelles **X** 46
Seymouria **F** 23
Shantungosaurus **F** 34
Sharks **C** 38, 39; **F** 19; **L** 40–41
Sharovipteryx **F** 40
Sheep **D** 52, 53
Shellfish **C** 18, **G** 54
Shells **F** 6, 8; **S** 9, 16
Ships **L** 52–53; **U** 9; **W** 39–43
Shockley, Walter **W** 10
Shockley, William **Q** 25
Sholes, Henry **Q** 11
Shops **R** 52–53
Shores
 animals **L** 19
 birds **B** 36; **L** 50
 homes in **L** 22–23
 muddy **L** 22–23
 plants **L** 19, 22
 rocky **L** 23
 sandy **L** 20
 silt **L** 22
 see also Beaches

KEY TO TITLES OF VOLUMES

A Plants
B Birds
C Cold-blooded Animals
D Warm-blooded Animals
E Insects and their Relatives
F The World Before Man
G The Beginnings of Life
H The Human Body
I Disease and Medicine
J The World Beneath Us
K The Invisible World
L Seas and Oceans
M Weather and its Work
N Astronomy
P Mathematics
Q Computers: An Introduction
R Working with Computers
S Through the Microscope
T How Everyday Things Work
U Great Discoveries and Inventions
V How Does it Work?
W Communications and Transport
X The World Today
Y Projects

Shrews
  common **D** 17
  elephant **D** 16
  otter **D** 16
  Savi's white-toothed **D** 16, 17
Shrikes **B** 31
Shrimps **A** 22, 51; **C** 28–30, 34; **E** 52, 53; **L** 33, 34
Shrubs **A** 22, 51
Sickle-cell anaemia **G** 58
Sierra Leone **X** 45
Sight **B** 10; **Y** 40, 41
Signals **Y** 16–17
Sikhs **X** 39
Sikorsky, Igor **V** 13; **W** 50, 57
Silica **J** 22
Silica cement **J** 21
Silicon-iron alloy **S** 34
Silk **C** 36–37
Silk moths **E** 18, 19
Silurian period **F** 14, 15
Silver **J** 30–31; **Y** 19
Silverfish **E** 20; **S** 25
Simpson, Dr Gaylord **D** 24
Simpson, James Young **I** 27
Simulations **R** 21–23
Singapore **X** 39
Singing insects **E** 28–29
Single cells **S** 18
Single-celled animals **C** 8–9
Siphonophores **C** 7
Sirius **N** 6
Sisal **G** 20
Sisters of Charity **I** 50
Skates **C** 38, 39
Skeleton
  bird **B** 7
  clues in **D** 10
  feet **D** 11
  human **H** 16
  vertebrae **D** 10
Skin
  animals **H** 46
  cells **S** 40, 41
  dermis **H** 46
  epidermis **H** 46
  follicle **H** 46
  function **H** 46–47
  keratin **H** 46
Skinks **C** 6, 54
Skull **H** 16
Skunk **D** 35
Sky **K** 20
Skylab **V** 52, 56, 58
Slate **J** 22
Slavs **X** 32
Sleep **K** 46–49
Sleeping sickness **S** 47
Sleet **M** 19
Slipher, Vesto Melvin **N** 48
Slipper limpet **G** 13
Sloth **D** 8, 11, 22, 23
Slow-worms **C** 54, 55
Slugs **C** 18, 20–21, 22; **F** 16; **G** 11, 28
Small life in water **S** 10–11
Smallpox **I** 17–18
Smelling **H** 42–43; **Y** 40, 42
Smith, William **F** 9; **J** 50
Smog **M** 33
Smoking **I** 60
Snails **C** 17, 18, 20–21, 22; **F** 16; **G** 11, 28
Snake millipede **C** 27
Snakebird **B** 24
Snakes **C** 56–57; **G** 38
Snappers **C** 41
Snorkel **V** 51
Snow
  air temperature **M** 18, 19
  crystals **M** 18–19; **P** 6
  flakes **M** 19
  layer **M** 19
  scooter **V** 17
Snowdrops **A** 29
Soap **U** 17
Sobrero, Ascanio **V** 42
Socotra **A** 61
Sodium **K** 36
Software
  birth of industry **Q** 38–40
  business **Q** 42
  early **Q** 25
  educational **Q** 43
  preparing for action **R** 9
  system **Q** 43
Softwood **A** 14
Sol *see* Sun
Solar furnace **K** 25
Solenodon **D** 17
Solid wheels **W** 18
Solids
  meaning **K** 6
  molecules **K** 16
  structure **K** 6
Solomon Islands **X** 51
Somalia **X** 45
Sommellier, Germain **W** 33
Songs and singing **B** 13, 54
Sonic booms **V** 10
Sony **W** 10
Sopwith Pup fighter **V** 8
Sound energy **K** 26
Sound track **T** 27
Sounds **W** 6
Source, The **R** 48
South Africa **X** 40, 46
South America **A** 21, 36; **X** 10, 20, 22–25
South Korea **X** 36
South Yemen **X** 38
South-eastern Asia **X** 39
South-western Asia **X** 9, 37–38
Southern Africa **X** 46
Southern Asia **X** 38–39
Southern Europe **X** 30
Soviet Union *see* Union of Soviet Socialist Republics
Space
  depths of **N** 47
  emptiness **K** 6
  exploration **V** 54–56
  flight **V** 52–59
  invaders **Q** 41; **R** 18, 19
  observatories **N** 60, 61
  probes **N** 60–61; **V** 56, 60
  recognizing stars and planets **Y** 22–23
  research **I** 61
  rockets **K** 14–15; **V** 19–20
  shuttle **W** 2–3, 5, 12
  stations **V** 56, 59
Spacecraft **Y** 14, 33
Spacewar **R** 18
Spadefoot toad **C** 47
Spain **X** 22, 30, 40
Spanner **V** 35
Sparrows **B** 6, 57
Spawning **G** 26–27, 34, 36, 37
Speak 'n' Spell **R** 40
Specialists, medical **I** 9, 49–50
Species **G** 55–56
Specific gravity **Y** 27
Spectacles **T** 35
Spectrum **N** 40
Speech
  discovery of **U** 6–7
  how we speak **H** 40–41
  language and **W** 6
  recognition **R** 43
  synthesis **R** 40, 41, 44, 50
  therapists **I** 54
Sperm **G** 11, 16, 26, 27, 34, 36, 49, 50, 53
Sphinx **N** 7
Spiderlings **C** 37; **G** 30
Spiders
  food **E** 56
  in the bath **S** 25
  internal fertilization **G** 28
  nets **E** 56–57
  silk **C** 36–37; **E** 56, 57
  spinnerets **E** 56
  webs **E** 56–57
Spinal cord **H** 32, 33, 34
Spines **D** 8
Spinetail **B** 47
Spiny anteater **D** 12, 13; **G** 44
Spiny cockle **C** 19
Spiny sharks **F** 19
'Spirit of St Louis' **V** 8
Spirogyra **S** 18, 22
Spitfire fighters **V** 8
Spleen **H** 27
Spoked wheels **W** 18
Sponges **C** 10–11, 13, 21; **G** 54; **L** 30; **S** 11
Spores **G** 16; **S** 31
Sporophytes **A** 12
Sporozoa **C** 8–9
Spotted cuscus **D** 14
Sprague, Frank **W** 29
Spreadsheets **Q** 39
Springtails **E** 20; **S** 29
Sprites **R** 30, 31
Spruces **A** 15
Sputnik **V** 52; **W** 12
Square root **P** 35
Squidlet **S** 10
Squids **C** 22; **G** 27
Squirrels **D** 27; **G** 23
Sri Lanka **X** 10, 35, 38
St Helena tortoise **C** 52
Stack pointer **Q** 12
Stag beetle **E** 48, 50
Stagecoaches **W** 19
Stalactites **J** 22, 49; **Y** 24–25

Stalagmites **J** 22, 49; **Y** 24–25
Starfish **C** 24, 25; **G** 8, 27; **L** 35
Starlings **B** 58
Stars
 birth and death **N** 15
 brightness **N** 40
 circumpolar **N** 36
 classification **N** 41
 clock **N** 7
 diameters **N** 41
 distances **N** 40
 double **N** 41
 eddies **J** 9
 emptiness **K** 6, 7
 gases **K** 8, 20
 globular star clusters **N** 43
 heat **N** 10
 names **N** 38
 nebulae **N** 44–46
 open star clusters **N** 43
 pulsars **N** 42–43
 recognizing **Y** 22
 super-heavy **K** 18
 temperature **N** 41
 variable **N** 41–42
 see also Constellations
Starvation **I** 41
Statement
 arithmetic **Q** 31, 32
 assignment **Q** 31
 command **Q** 31
 input **Q** 31
 output **Q** 31
Statistics
 average **P** 48
 collecting data **P** 48–49
 earliest **P** 48
 examples of misuse **P** 47–48
 mean **P** 48
 median **P** 48
 mode **P** 48
 probability theory **P** 40, 50
 processing data **P** 49
 sample **P** 48
 statistical distribution **P** 47
Steam
 age of **W** 26–28
 engines **V** 28–29
 heating **T** 25
 plough **U** 11
Steel **Y** 32
Stegosaurus **F** 30, 36
Stems **A** 40
Stenonychosaurus **F** 31
Stentor **S** 14
Stephenson, George **V** 29; **W** 26–27, 33
Stephenson, Robert **W** 33
Sterile clothes **I** 57
Sterilization **I** 23
Stevenson screen **M** 31
Stickleback **G** 36
Stinging cell **C** 12
Stinging nettles **A** 25
Stink glands **C** 27
Stoat **D** 34
Stomach
 acid juices **H** 24
 contractions **H** 24
 digestive juices **H** 24
 mucus **H** 24
 structure **H** 26
Stone
 Age **J** 30; **U** 6
 axes **U** 6
 flanges **U** 6
 tools **U** 6
Stonehenge **N** 8, 14
Storage in computers
 permanent **Q** 7
 short-term **Q** 7
Storks **B** 35
Strassman, K **U** 54
Strata **J** 6, 7, 15
Strawberries **A** 34; **G** 24
Street cross-section **W** 15
Streptococci **S** 47
String telephone **Y** 16
Stromatolites **F** 17
Struthiosaurus **F** 36
Stuart, Herbert Akroyd **V** 32
Students **I** 49
Stylonychia **S** 15
Submarines **V** 50–51; **W** 39–41; **Y** 34
Submersibles **L** 55, 56; **V** 51
Substances
 changing **K** 34
 chemical **K** 36
 radioactive **K** 23
 super-heavy **K** 18
Subtraction **P** 12
Subungulates **F** 53–54
Sudan **X** 44
Suez Canal **L** 54
Sugar **K** 16, 26, 37–39
Sugar beet **A** 39
Sugar cane **A** 38, 39
Sugar tongs **V** 33
Sugarloaf mountain, Brazil **J** 15
Sulphur **J** 39; **S** 32
Sumerians **A** 48; **W** 7, 18, 19; **Y** 8

THE SIZE OF IT

The Earth could swallow up 17 planets the size of the tiny planet Mercury; but Jupiter, largest of the planets, could swallow up 1300 Earths and still have room for more. And 1,000 Jupiters could fit into the Sun. But the Sun is very far from being the largest star – the red giant Betelgeuse, in the constellation Orion, could contain 500 million stars the size of the Sun.

Sun
 birth of **N** 14
 death of **N** 14
 distance from Earth **N** 12
 eclipse **N** 11, 24
 effect on tides **L** 10
 energy from **N** 12–13
 family **N** 10
 gases **K** 20
 heat **N** 13
 hydrogen **K** 9
 light from **N** 13
 projecting image from **N** 15
 prominences **N** 13, 14
 radiation **Y** 37
 rays **K** 10–11
 revolving of **N** 8, 9
 section **N** 12
 solar furnace **K** 25
 solar wind **N** 13
 sunspots **N** 13–14
 temperature **N** 13
Sunbirds **B** 55
Sundews **S** 8
Sundial **Y** 29
Sungrazer comets **N** 32
Suni **D** 58
Supermarkets **R** 52, 53
Supersonic flight **V** 10
Surface tension **Y** 27
Surgeons **I** 49, 56–57
Surgery **I** 56–57
Surgical instruments **I** 56
Surinam **X** 22
Surinam toad **C** 47
Surveyor **J** 11
Sutton, Henry **Y** 29
Swallow **B** 43, 51
Swallowtail butterfly **E** 35
Swans **B** 21, 26, 28
Swaziland **X** 46
Sweat **H** 46
Sweden **X** 27, 28
Sweetlips **C** 41
Swifts **B** 9, 43; **C** 58, 59
Switch system **R** 51
Switches **Q** 7
Switzerland **X** 30
Swordfish **C** 40
Symbols
 acceptance of **P** 35
 equals **P** 14
 larger than **P** 14
 numbers as **P** 10
 smaller than **P** 14
 working with **P** 14–15
Symmetry
 bilateral **P** 21
 in humans **P** 21
 introduction to **P** 6
 meaning **P** 19–20
 plane of **P** 21
 practical **P** 21
 rotational **P** 21
Symptoms **I** 9
Synapse **H** 33
Synclavier **R** 36
Synth-Axe **R** 37
Synthesizers **R** 36–37, 40, 41, 44, 50
Synthetic detergents **U** 17
Syracuse, Sicily **Y** 27
Syria **W** 7; **X** 37, 38

GRAINS OF TRUTH [*Double trouble!*]

A king once asked his court mathematician what reward he would like to receive for his services. The wily mathematician replied that he would like to receive some wheat – as many grains as it would take to place one grain on the first square of a chessboard, two grains on the second square, four on the next, eight on the next, and so on, doubling the number each time until the 64th and final square was reached. (Can you guess now how many grains would be needed?) The king rashly promised to do this – and then found that he would need more wheat than all the granaries of the world could ever hold. In fact, it would take *over 18 million million million* grains of wheat (which is millions of millions of tonnes) to give the mathematician what he asked for.

T

▲ Three telescopes at an observatory in Italy. When not in use, the dome of the observatory swivels shut to protect them.

Tables
  trigonometric **P** 23
Tablets
  digitizing **Q** 58; **R** 32
  graphics **R** 32
Tadpoles **C** 46; **G** 11; **S** 10
Tailorbird **B** 13
Taiwan **X** 36
Talkative robots **R** 60
Talking Calculator **R** 41
Talking computers **R** 40
Talking Story **R** 41
Talons **B** 31
Tamandua **D** 22
Tanystrophaeus **F** 29
Tanzania **X** 40, 45–46
Tape recorder
  cassette **T** 48
  playback **T** 48
  recording **T** 48
  stereophonic **T** 48
Tapeworms **G** 11
Tapirs **D** 43
Tasmania **X** 48
Tasmanian
  wolf **D** 14
Taste
  birds **B** 11
  buds **H** 42
  human **H** 42–43
  insects **E** 13
  testing **Y** 40, 41
Teaching mathematics **P** 60
Teeth
  bridge **I** 59
  caries **I** 58–59
  crown **I** 59

decay **I** 58–59
description **D** 11
elephants' **D** 40
false **I** 59
hares' **D** 24
insectivores' **D** 16
lower **I** 58
rabbits' **D** 24
rodents' **D** 27
shape **D** 11
upper **I** 58
wild dogs' **D** 29
Tele-typewriter **W** 8
Telegraph **W** 8
Telepathy **K** 58
Telephone exchange **W** 9, 13
Telephones **R** 44–45, 50; **V** 26, 27, 47–48; **W** 8, 9, 13; **Y** 16
Telescopes
construction **Y** 32
developments **N** 54–55
dish radio **N** 59
Earl of Rosse's **N** 46, 55, 56
early mathematicians **P** 24–25
linear radio **N** 60
modern **N** 56; **Y** 23
origins **N** 52–54; **U** 10
radio **N** 56–59
recognizing stars and planets **Y** 22
William Herschel's **N** 55
Teletext **W** 13
Television
camera **T** 58
colour **T** 59–60
communication **W** 10–11
computer uses **Q** 6
electron gun **T** 58
electrons **T** 58–59
Telex **W** 8, 9
Telford, Thomas **W** 14
Telling the Time **R** 41
Telstar **W** 12
Temperate climates
cool **M** 46–47
location **M** 46
warm **M** 47
Tenontosaurus **F** 34
Teratornis **F** 48
Terminator **Q** 35
Termites
diet **E** 32
homes **E** 32
king **E** 33
queens **E** 33
soldiers **E** 33
workers **E** 32–33
Terns **B** 37
Terrapins **C** 51, 52
Terrigenous deposits **L** 15
Tertiary period **F** 46
Tests
chemical **Y** 24–25
computers **Q** 35
electricity **Y** 30–31
magnetism **Y** 28–29
physical **Y** 26–27
senses **Y** 40–43
Tetraploidy **G** 58
Texas Instruments **R** 40; **W** 10
Text by telephone **R** 44, 45
Textiles
early textile-making **U** 27
loom **U** 27
mill **U** 27
thread **U** 27
Thailand **X** 39
Thecodonts **F** 28, 31
Theodolites **P** 44
Theophrastus **N** 13
Theorems **P** 38
Therapsids **F** 25
Therapy **I** 54–55
Theriodonts **F** 25
Thermals **V** 14
Thermionic valve **W** 10
Thermographs **H** 60; **I** 46–47; **M** 31
Thermometers
bimetallic strip **T** 29
celsius scale **T** 27
construction **T** 27
fahrenheit scale **T** 27
industrial **T** 29
invention **U** 11
liquid-in-glass **T** 27

▼ The Mont Cenis Tunnel was the first of the tunnels through the Alps linking France and Italy.

KEY TO TITLES OF VOLUMES

A Plants
B Birds
C Cold-blooded Animals
D Warm-blooded Animals
E Insects and their Relatives
F The World Before Man
G The Beginnings of Life
H The Human Body
I Disease and Medicine
J The World Beneath Us
K The Invisible World
L Seas and Oceans
M Weather and its Work
N Astronomy
P Mathematics
Q Computers: An Introduction
R Working with Computers
S Through the Microscope
T How Everyday Things Work
U Great Discoveries and Inventions
V How Does it Work?
W Communications and Transport
X The World Today
Y Projects

LONGEST BRIDGES

These bridges are longer than 700 m (2,300 ft). They are all suspension bridges: that is, the part of the bridge over which the traffic goes is suspended from cables which pass over tall towers near each end of the bridge and are firmly anchored at either end.

Length (m)	Length (ft)	Name	Year of completion	Location
1,780	5,840	Akashi-Kaikyo*	1988	Honshu-Shikoku, Japan
1,410	4,626	Humber Estuary Bridge	1980	Humber, England
1,298	4,260	Verrazano-Narrows (6+6 lanes)	1964	Brooklyn-Staten I, USA
1,280	4,200	Golden Gate	1937	San Francisco Bay, USA
1,158	3,800	Mackinac Straits	1957	Straits of Mackinac, Mich, USA
1,074	3,524	Bosphorus Bridge	1973	Bosphorus, Istanbul, Turkey
1,067	3,500	George Washington (2 decks, lower has 6 lanes, upper has 8 lanes since 1962)	1931	Hudson River, NY City, USA
1,013	3,323	Ponte 25 Abril (Tagus) (4 lanes)	1966	Lisbon, Portugal
1,006	3,300	Firth of Forth Road Bridge	1964	Firth of Forth, Scotland
988	3,240	Severn-Wye River (4 lanes)	1966	Severn Estuary, England
876	2,874	Ohnaruto	1983	Kobe-Naruto Route, Japan
853	2,800	Tacoma Narrows II	1952	Washington, USA
770	2,526	Innoshima	1982	Onomichi-Imapari Route, Japan
712	2,336	Angostura	1967	Ciudad Bolívar, Venezuela
712	2,336	Kanmon Straits	1973	Shimonoseki, Japan
704	2,310	Transbay (2 spans)	1936	San Francisco-Oakland, Calif, USA
701	2,300	Bronx-Whitestone (Belt Parkway)	1939	East River, NY City, USA

*due to open in 1988

KEY TO TITLES OF VOLUMES

A Plants
B Birds
C Cold-blooded Animals
D Warm-blooded Animals
E Insects and their Relatives
F The World Before Man
G The Beginnings of Life
H The Human Body
I Disease and Medicine
J The World Beneath Us
K The Invisible World
L Seas and Oceans
M Weather and its Work
N Astronomy
P Mathematics
Q Computers: An Introduction
R Working with Computers
S Through the Microscope
T How Everyday Things Work
U Great Discoveries and Inventions
V How Does it Work?
W Communications and Transport
X The World Today
Y Projects

medical **T** 27, 28
resistance **T** 29
thermocouple **T** 29
types **T** 27, 29
ways of marking temperatures **T** 27
Theropods **F** 30, 31
Thomson, Joseph **U** 12
Thorntail **B** 44
Thoughts **K** 44–45
Thrasher **B** 53
3-D graphics **R** 21
Thrinaxodon **F** 29
Throat **S** 38, 47
Thrushes **B** 11, 54
Thunder
clouds **M** 11, 16, 17
sounds energy **M** 17
storms **M** 16–17
Thylacine **G** 60
Thyroid gland **H** 28
Tibetans **X** 35, 36
Tides
ebbing **L** 20
effect on Moon **N** 23
gravity **N** 17
neap **L** 10
range **L** 10
spring **L** 10
working **L** 10
Tiger **D** 36, 37
Time **P** 45–46
Time chart **J** 63
Tin **Y** 32
Tissue-typing **I** 58
Tissues **H** 12; **S** 42–43
Tit **B** 13
Titan **N** 27
TNT **V** 42
Toads **C** 46, 47; **G** 37; **Y** 53
Toadstools **S** 31
Tobacco **G** 20
Tobago **X** 19
Todies **B** 45
Togo **X** 45
Tokyo **K** 27
Toltec **X** 13
Tombaugh, Clyde **N** 29
Tonatiuh **Y** 12
Tonga **X** 51
Tonga tortoise **C** 52
Tongue **H** 42; *see also* Taste
Tonsilitis **H** 22
Tonsillotome **H** 23
Tonsils **H** 22, 23
Tools
metal **U** 8
stone **U** 6
Toothless mammals **D** 22–23
Topology **P** 41–42
Tornadoes
countries suffering from **M** 25
damage caused by **M** 25
waterspouts **M** 25
winds **M** 25
Tortoises **C** 51–52; **F** 29; **G** 38, 39, 59
Tortoiseshell butterfly **E** 35
Toucans **G** 23
Touch
insect **E** 13
testing **Y** 40
Tour de France **T** 8
Townes, Charles **U** 58
Towns, development of **U** 8
Track ball controller **Q** 6
Trade **U** 9
Trade winds **M** 20, 21
Trains **W** 29–31
Tranquilizer **I** 59
Transatlantic telephone service **W** 8
Transistor
computer uses **Q** 7, 25
development **W** 10
diodes **T** 54, 56
electrons **T** 54
invention **U** 56
radio **T** 54
semiconductors **T** 54
working of **T** 56–57
Transplants
bone **I** 58
cornea **I** 58
heart **I** 57, 58, 60
kidneys **H** 27; **I** 57–58
rejection **I** 57
skin **I** 57
Transport (volume **W**)
air **W** 46–61
future of **W** 61
road **W** 14–25
trade and **U** 9
water **W** 34–35
Traumatic illness **I** 6
Travelators **T** 16
Travelling energy **K** 23, 24
Trawlers **L** 58, 59
Treatments, names for **I** 6–7
Trees **K** 37
Trees with cones and needles **A** 14–15
Trematosaurus **F** 29
Trésuaget, Pierre **W** 14
Trevithick, Richard **W** 26, 27
Triangles
equilateral **P** 16
isosceles **P** 20
properties **P** 22
right-angled **P** 22
working with **P** 22–23
Triassic times **F** 26–28
Triceratops **F** 37
Trick drawing and writing **Y** 8–9
Tricycle **T** 6
Trigonometry
calculations using **P** 22–23
higher **P** 41
sailors' use of **P** 22
tables **P** 23
Trilobites **F** 14, 15, 16
Trinidad **X** 19
Trinitrotoluene **V** 42
Triton **N** 28; **V** 50
Tropical
fish **Y** 59
grassland **M** 42, 43
rainforests **A** 21–23, 60
rainy climates **M** 38–39
Tropics **A** 43–45
Troposphere **M** 6
Trunk **A** 27
Ts'ai Lun **U** 23
Tsetse fly **E** 16, 38
Tsiolkovsky, Konstantin **U** 51
Tsunamis **L** 10
Tswett, Mikhail **U** 16
Tuatara **C** 48
Tuberculosis **I** 28
Tubers **G** 24–25
Tulips **A** 18, 19, 31, 50, 51
Tulp, Nicholaas **H** 10
Tumble polishing **Y** 18
Tuna fish **C** 40
Tundra **A** 6, 12
Tunisia **X** 44
Tunnels **W** 16, 17, 33
Turbine
Francis reaction **U** 45
Kaplan **U** 45
Pelton wheel **U** 45
working of **V** 40–41
Turkey
bird **B** 33
people of **X** 26, 34, 37, 38
Turkey, brush **B** 34
Turkic peoples **X** 35
Turtle dove **B** 38
Turtles **C** 51–53, 62; **F** 38; **G** 38, 39 **L** 50–51
Turtles in classroom **R** 60
Tuvalu **X** 51
TV games **Q** 40–41
Twisters **M** 25
Two-part life cycles **G** 16
Two-winged flies **E** 38–39
TWX machines **W** 8
Tycho *see* Brahe, Tycho
Tylosaurus **F** 38
Typewriter
carriage **T** 20
dot matrix printing **T** 21
electric **T** 20
electronic **T** 20–21
golfball **T** 20, 21
manual **T** 20
musical scores, for typing **T** 20
Remington **T** 20
ribbon **T** 20
word processor **T** 21
Typhoid **I** 29
Typhoons **M** 24
Typhus **I** 32–33
Tyrannosaurus **F** 30, 31, 37, 42
Tyres
burnout **T** 13
casing **T** 12
construction **T** 12
cross ply **T** 12
development **W** 18
early **T** 12; **U** 36, 37
pneumatic **T** 12
radial ply **T** 12
sizes **T** 12
solid rubber **T** 12
tubeless **T** 12
uses **T** 12
vulcanization **T** 12
worn **T** 13

▲ USA: San Francisco is in a region where earthquakes occur. The Trans-America building in the city has been designed in such a way that it is believed to be earthquake-proof.

US Geological Survey **Y** 21
Uighurs **X** 36
Ulothrix **S** 18
Ultrasonics **R** 60
Ultrasound **H** 8; **I** 43
Ultraviolet rays **K** 11
Understorey plants **A** 22
Ungulates **F** 52–53
Unidentified flying objects **K** 56–57
Union of Soviet Socialist Republics (USSR) **A** 15; **X** 13, 26 28, 31, 32–33, 34, 36, 38
United Arab Emirates **X** 37
United States of America **X** 10, 13, 14, 15–16, 18, 22, 32, 33
Universal gravitation **N** 16–17
Universe
  atoms and **K** 28–29
  emptiness **K** 6, 7
  expanding **N** 48
  origin of **N** 50–52
Universities **I** 14, 49
Upper Volta *see* Burkina Faso
Uranium **V** 36–38
Uranus
  diameter **N** 28
  discovery **N** 10
  moons **N** 28
  orbit **N** 11
  rings **N** 28
  rotation **N** 28
Urine **H** 25
Uruguay **X** 22
Useful insects **E** 18–19
User friendliness **R** 16
User port **Q** 16
Using computers
  home computer equipment **R** 9
  other hardware **R** 9

▲ The volcano Haleakala on Maui on Hawaii has the largest crater in the world.

Vaccination **I** 17–18, 21
Vacuum tube **W** 10
Valhalla **R** 20
Valves **W** 10
Van Gogh, Vincent **I** 39
Van Leewenhoek, Anton *see* Leewenhoek, Anton van
Van de Poele **W** 29
Vane, feather
  aftershaft **B** 6
  barbs **B** 6
Vanuatu **X** 51
Vapour **K** 6
Variable suffixes **Q** 29
Variables **Q** 28, 29, 30, 33–34
Vatican City **X** 31
Veddoids **X** 10, 35
Vegetables **A** 40–41
Vegetation **A** 6–7, 18–19
Vegetative reproduction **G** 24–25
Veins **H** 11, 18, 19
Venezuela **X** 22, 23
Venus
  atmosphere **N** 19
  diameter **N** 19
  discovery of **N** 10
  distance from Sun **N** 19
  flytrap **A** 55
  phase effects **N** 19
  probes **V** 56
  surface **N** 19
Vertebrae **H** 33
Vertebrates **G** 47
Vertical take-off **W** 56–57
Vesalius, Andreas **H** 8, 10; **I** 14, 15
Vesta **N** 30
Vicuña **D** 49
Video
  cassette **T** 61
  games **T** 61
  recording **T** 61
  video-disc playback **T** 61
Video RAM **Q** 11
Videotext **R** 44
Viete, Francois **P** 35
Vietnam **X** 39
Viewdata **W** 13
Viewdata service **R** 47, 49
Viking longships **W** 34–36
Viking probes **N** 25; **V** 56
Villi **H** 25
Vinci, Leonardo da **H** 10; **U** 10; **W** 44
Vinegar **K** 36
Vines **A** 46
Violets **A** 18
Virologist **S** 59
Viruses **H** 25; **I** 16, 19–21, 30–32; **S** 46, 58–59
Visions **K** 56–57
Visual display unit **Q** 9
Visual signals **W** 8
Vitamins **A** 40, 41; **H** 51; **I** 41; **K** 40, 41
Vizaspell **R** 25
Vizawrite **R** 25
Vocabulary of words **W** 6
Vocal cords **H** 40; **W** 6
Voice box **W** 6
Voicemate **R** 43
Voisin, Gabriel **W** 48
Volcanoes
  different types **J** 26
  earth formation **J** 8
  eruption **J** 26, **K** 33
  extinct **J** 26
  geology **J** 14
  Hawaiian **J** 26
  lava **J** 25
  meaning **J** 25
  Pelean **J** 26
  Plinian **J** 26
  rocks **J** 19
  South American **K** 20–21
  Strombolian **J** 26
  vent **J** 25
  Vulcanian **J** 26
Volk, Magnus **W** 29
Volvox **G** 14
Volvox colonies **C** 8
von Leibnitz, Gottfried *see* Leibnitz, Gottfried von
von Ohain, Dr Hans *see* Ohain, Dr Hans von
Vowel sounds **W** 6
Voyager probe **N** 60; **V** 56, 57
Vultures **B** 9, 15, 20

KEY TO TITLES OF VOLUMES

A Plants
B Birds
C Cold-blooded Animals
D Warm-blooded Animals
E Insects and their Relatives
F The World Before Man
G The Beginnings of Life
H The Human Body
I Disease and Medicine
J The World Beneath Us
K The Invisible World
L Seas and Oceans
M Weather and its Work
N Astronomy
P Mathematics
Q Computers: An Introduction
R Working with Computers
S Through the Microscope
T How Everyday Things Work
U Great Discoveries and Inventions
V How Does it Work?
W Communications and Transport
X The World Today
Y Projects

► The waterfalls at Iguazú on the borders between Brazil and Argentina.

W

KEY TO TITLES OF VOLUMES

A Plants
B Birds
C Cold-blooded Animals
D Warm-blooded Animals
E Insects and their Relatives
F The World Before Man
G The Beginnings of Life
H The Human Body
I Disease and Medicine
J The World Beneath Us
K The Invisible World
L Seas and Oceans
M Weather and its Work
N Astronomy
P Mathematics
Q Computers: An Introduction
R Working with Computers
S Through the Microscope
T How Everyday Things Work
U Great Discoveries and Inventions
V How Does it Work?
W Communications and Transport
X The World Today
Y Projects

▼ Water vapour often gives rise to dramatic cloud formations, including these photographed in the Antarctic.

Waders **B** 36
Wagons **W** 18–19
Wallabies **D** 14; **G** 60
Walruses **D** 39; **F** 56, 57
Walton, Ernest **P** 43
Wapiti **D** 50
War control room **W** 13
Warfarin **G** 57
Warm-blooded animals *see* Mammals
Warm-blooded creatures **C** 6, 7
Wasps
hornet **E** 45
nests **E** 44
social **E** 44
solitary **E** 44–45
Watch
analogue **T** 52
clockwork **T** 52–53
digital **T** 52
electric **T** 52
quartz **T** 52, 53
spring **T** 52
Water
atoms **K** 24
beetles **E** 10; **Y** 53
boiling **K** 14, 16
breathing under **E** 10
cycle **K** 12–13
fleas **E** 52; **G** 13; **S** 23, 63
lily **A** 25; **G** 23; **Y** 53
mill **V** 44–45
plant's use of **A** 26–27
small life in **S** 10–11
solid **K** 16
spiders **C** 36
turbine **V** 41
underground **J** 47–49
vapour **M** 7
Waterfalls **U** 44–45
Watson, Dr **Y** 42
Watt, James **V** 28
Wavellite **K** 17
Waves
body **K** 45
brain **K** 44–45
crest **L** 9, 10
effect of wind **L** 10
energy **K** 22–23
heat **K** 22–23
height of ocean **L** 10
length of ocean **L** 10
light **K** 22–23
mind **K** 58
ray **K** 23
spindrift **L** 9
storm **L** 10
tsunamis **L** 10

wind-driven **L** 10
see also Rays
Wayne, John **I** 45
Weasels **D** 34–35
Weather (volume **M**)
altering **M** 32
altering by accident **M** 32–33
at work **J** 23–24
forecasting *see* Weather forecasting
ingredients in **M** 6
Weather forecasting
by swifts **B** 43
computer uses **Q** 45
instruments for measuring **M** 30–31
making a forecast **M** 31
radiosonde **M** 30
short-range **M** 31
Weaverbirds **B** 57
Weavers **B** 57
Web spiders **C** 36
Weevil **E** 48, 49, 50
Weigeltisaurus **F** 40
Weight
effect of gravity on **N** 17
Weightlessness **V** 56, 58–59
Weights and measures
British imperial **P** 44
metric **P** 44
SI system **P** 45
standard kilo weight **P** 44
US customary **P** 44
Werner, Eugene **V** 16
Werner, Michael **V** 16
West Africa **X** 45
West Germany **X** 28, 30
West Indies **X** 14, 19–20
West Pakistan **X** 38
Westerlies **M** 21, 22
Western Samoa **X** 51
Westinghouse Corporation **R** 60
Whales
baleen **L** 48–49; **Y** 54
blue **D** 60–61
body **L** 48
conservation **L** 60–61
early **F** 56
killer **D** 7
shark **C** 38
skeleton **D** 10
skin **G** 44
sperm **L** 5
toothed **D** 60; **L** 5, 48
whalebone **D** 60
Wheatstone, Charles **W** 8
Wheels
gears **U** 37
origins **U** 36–37
overbalancing **V** 45
solid **W** 18
spinning **U** 36
spoked **U** 37; **W** 18
Whelks **C** 21
Whip-poor-will **B** 42; **C** 58
White butterflies **E** 34
White mice **D** 26
White rats **D** 26
Whittle, Sir Frank **W** 53
Wild asses **D** 42–43
Wild boar **D** 47
Wild cattle **D** 56–57
Wild plants **A** 60–61
Wilkes, Maurice **Q** 22
Williams, F C **Q** 22
Wind-borne seeds **G** 22
Windlass **V** 33
Windmills
pumps **T** 14
working **V** 40
Winds
anabatic **M** 53
anticyclones **M** 22
attacks on land **M** 60
changing direction **M** 20
cyclones **M** 22
desert **M** 45
Doldrums **M** 20
earth's rotation **M** 21
effect on waves **L** 10
great wind systems **M** 20–21
Horse Latitudes **M** 20
hurricane **M** 24
katabatic **M** 53
mountain **M** 52–53
names of **M** 20
Polar Easterlies **M** 21, 22
sand blown by **M** 60, 61
strong **M** 21
tornadoes **M** 25
trade **M** 20, 21
Westerlies **M** 21, 22
Wine **A** 46; **S** 37
Wine making **Y** 46–47
Wingless insects **E** 20–21
Winkles **C** 21; **G** 13, 28
Wireworms **E** 50
Wobbegong **C** 39
Wolf spiders **E** 56, 57
Wolverine **D** 34, 35
Wolves **D** 28, 29; **G** 56, 60

The tallest buildings in the world are TV and radio transmitter masts, which have to be high to reach the greatest possible audience. These are listed below, together with the tallest office buildings.

WORLD'S TALLEST STRUCTURES

Height (m)	(ft)	Structure	Location
646	2,063	Warszawa Radio Mast (May 1974)	Konstantynow, nr Płock, Poland
629	2,120	KTHI-TV (December 1963)	Fargo, North Dakota, USA
553	1,815	CN Tower, Metro Centre (April 1975)	Toronto, Canada
537	1,762	Ostankino TV Tower (1967) (4 m added in 1973)	near Moscow, USSR
533	1,749	WRBL-TV & WTVM (May 1962)	Columbus, Georgia, USA
533	1,749	WBIR-TV (September 1963)	Knoxville, Tennessee, USA
528	1,732	Moscow TV Tower	Moscow, USSR
510	1,673	KFVS-TV (June 1960)	Cape Girardeau, Missouri, USA
499	1,638	WPSD-TV	Paducah, Kentucky, USA
493	1,619	WGAN-TV (September 1959)	Portland, Maine, USA
490	1,610	KSWS-TV (December 1956)	Roswell, New Mexico, USA
487	1,600	WKY-TV	Oklahoma City, Okla, USA
479	1,572	KW-TV (November 1954)	Oklahoma City, Okla, USA
465	1,527	BREN Tower (unshielded atomic reactor) (April 1962)	Nevada, USA

THE WORLD'S TALLEST OFFICE BUILDINGS

Height (m)	(ft)	No of storeys	Building	Location
443	1,454	110	Sears Tower (1974)	Wacker Drive, Chicago, Illinois
412	1,350	110	World Trade Center (1973)[1]	Barclay and Liberty Sts, New York City
381	1,250	102	Empire State Building (1930)[2]	5th Av and 34th St, New York City
346	1,136	80	Standard Oil Building (1973)	Chicago, Illinois
343	1,127	100	John Hancock Center (1968)	Chicago, Illinois
320	1,049	75	Texas Commerce Plaza	Houston, Texas
319	1,046	77	Chrysler Building (1930)	Lexington Av and 42nd St, New York City
289	950	66	American International Building	70 Pine St, New York City
285	935	72	First Bank Tower	First Canadian Place, Toronto, Ontario
282	927	71	40 Wall Tower	New York City
279	914	59	Citicorp Center (1977)	New York City
262	859	74	Water Tower Plaza (1975)	Chicago, Illinois
261	858	62	United California Bank (1974)	Los Angeles, California
259	851	66	United California Bank	New York City

[1] Two TV antennae bring the overall height to 475·18 m, 1,559 ft. [2] Between 27 July 1950 and 1 May 1951 a 222 ft TV tower was added.

KEY TO TITLES OF VOLUMES

A Plants
B Birds
C Cold-blooded Animals
D Warm-blooded Animals
E Insects and their Relatives
F The World Before Man
G The Beginnings of Life
H The Human Body
I Disease and Medicine
J The World Beneath Us
K The Invisible World
L Seas and Oceans
M Weather and its Work
N Astronomy
P Mathematics
Q Computers: An Introduction
R Working with Computers
S Through the Microscope
T How Everyday Things Work
U Great Discoveries and Inventions
V How Does it Work?
W Communications and Transport
X The World Today
Y Projects

Womb **G** 47, 53
Wombat **D** 14
Wood **K** 42
Wood ant **E** 46
Wood's alloy **Y** 33
Woodchuck **C** 61
Woodpecker **B** 6, 40, 46
Woodpecker-finch **G** 60
Wool **D** 8
Word processing **Q** 38, 40; **R** 9, 24–26
Word processor **T** 21
Word stores **R** 42
Words **W** 6
Worker honeybee **C** 33
Working under water **L** 56–57
World War 2 **W** 52
World Wildlife Fund **A** 61
Worms
bristle **L** 32
earth **C** 14, 17, 34
evolution **G** 54
fan **L** 32
flat **L** 32, 33
marine bristle **C** 15–16; **G** 27
parthenogenesis **G** 13
round **L** 32, 33
sea **S** 13
tube-dwelling **C** 16
Wrench **V** 35
Wrens **B** 52
Wright, Orville **V** 6, 7, 12; **W** 48
Wright, Wilbur **V** 6, 7, 12; **W** 48, 49
Writing
and printing **W** 6–7
mirror **Y** 9
trick **Y** 8–9
Wyoming, USA **J** 14

► Waves are a tremendous source of energy, whether they are waves of heat, light or the sea.

XYZ

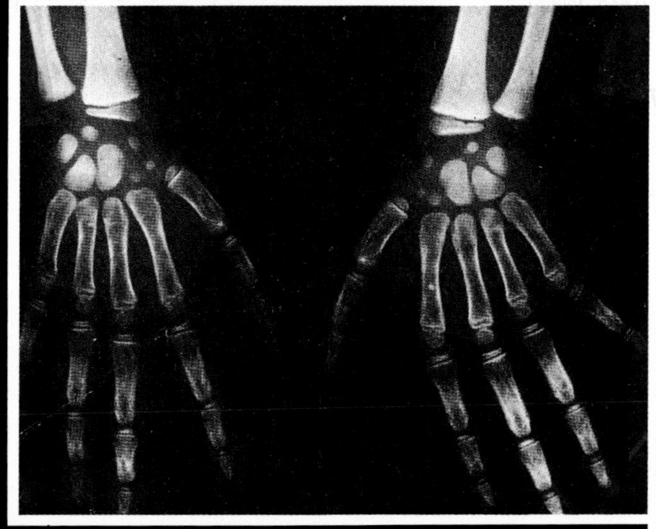

▲ X-rays enable doctors to see the bones of our bodies.

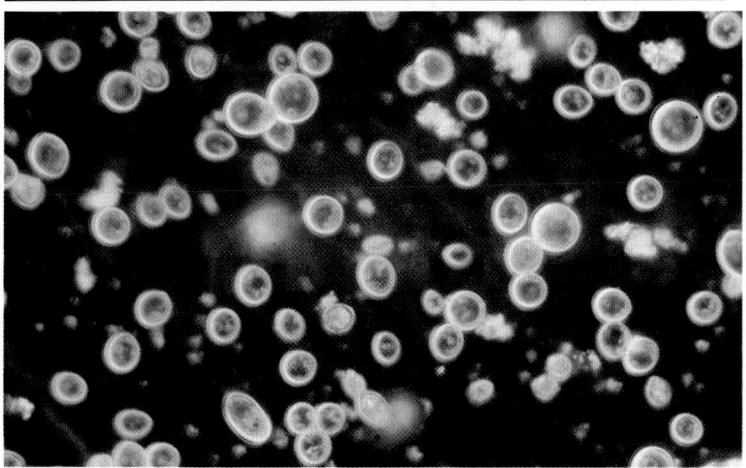

► Yeast cells are microscopic. These have been magnified 1000 times.

▲ A zebra mare with her new foal. Zebras live in herds and rely on their good hearing and speed for safety and young zebras are ready to join the herd within hours of their birth.

X-ray
- Coolidge tube **U** 46
- Crookes' tube **U** 46
- dental **U** 47
- discovery **U** 46
- energy **K** 24
- luggage checks **U** 47
- medicine **H** 59, 60; **I** 8, 28, 59
- radioactivity **K** 23
- using **U** 46
- what are **U** 46

Yachts **W** 38
Yak **D** 57
Yale, Linus **T** 19
Yamaha CX5M **R** 35
Yeast **A** 10; **S** 37; **Y** 46
Yellow fever **S** 58
Yemen Arab Republic **X** 38
Yew **A** 15, 35
Yoke **Q** 15
Yugoslavia **X** 31

Zaire **X** 45
Zambia **X** 46
Zapping aliens **P** 53; **R** 19
Zeaker **R** 54
Zebras **D** 42, 43
Zebu **D** 56
Zimbabwe **X** 46
Zinc **K** 16
Zodiac **N** 11
Zoos **S** 22–24
Zworykin, Vladimir **W** 10
Zygote **G** 11, 14, 16, 49

BINARY MATHS

To a computer, 10 + 10 = 100! Computers work with the binary system, which instead of the ten digits 0, 1, 2, 3, . . . 9 just uses the two digits 0 and 1. In our ordinary 'base-10' number system, a digit has 10 times its value if it's in the second column from the right, 100 times its value if it's in the third column, and so on. In the binary system, a 1 takes double, four times, eight times its value, and so on, as you go farther to the left. For example, 11111 represents (reading from the right): 1 + 2 + 4 + 8 + 16 – which is 31. And 10101 represents: 1 + 0 + 4 + 0 + 16 – that is, 21. So the numbers from 1 to 15 look like this:

base 10	binary
1	1
2	10
3	11
4	100
5	101
6	110
7	111
8	1000
9	1001
10	1010
11	1011
12	1100
13	1101
14	1110
15	1111

So what does an addition sum look like? Here are some examples (remember, when you add two 1s, you have to 'carry 1' to the next column):

```
base 10   binary

   2        10
 + 2      + 10
 ---      ----
   4       100

   3        11
 + 2      + 10
 ---      ----
   5       101

   7       111
 + 7     + 111
 ---     -----
  14      1110
```

What about multiplication? Since you only have to know the zero-times table and the 1-times table, it's pretty easy!

```
base 10   binary

   2         10
 × 2       × 10
 ---       ----
   4         00
             10
           ----
            100

   3         11
 × 2       × 10
 ---       ----
   6         00
             11
           ----
            110

   7        111
 × 5      × 101
 ---      -----
  35        111
            000
            111
         ------
         100011
```

PACKED WITH INFORMATION

All the information about how a human being should develop from an embryo to adulthood is contained in molecules of a substance called DNA, found at the centre of each body cell. The longest molecule of DNA in a human cell is about a hundredth of a millimetre (less than a two-thousandth of an inch) long. This is gigantic by molecular standards, but the molecule is actually tightly coiled, like a light bulb filament. If uncoiled, it would be 7000 times longer – 7.3 cm (3 in). The average person contains 100 g (3.5 oz) of DNA. If it were all uncoiled and laid end to end, it would stretch from the Earth to the sun and back 100 times.

THE LONGEST JOURNEY EVER

Two unmanned space probes, Voyager 1 and Voyager 2, will soon leave the solar system on journeys that will never end – unless they are captured by some other star in the Galaxy thousands of years from now. Voyager 2 was launched first, on 20 August 1977, and flew by Jupiter less than a year later. The giant planet's gravity hurled the craft on towards Saturn, which it passed on 26 August 1981. Voyager 2 passed Uranus on 24 January 1986 and will encounter Neptune, 4497 million km (2794 million mi) from the sun, on 24 August 1989. Voyager 1 was launched 16 days after Voyager 2, but reached Jupiter four months sooner, and Saturn nine months sooner. That was the last planet Voyager 1 encountered: it will leave the solar system in late 1988, while Voyager 2 will do so in 1990. From then on, both will be in interstellar space, and we shall never know their fate – unless an alien civilization reads the messages engraved on plaques on the probes' hulls, and sends a reply. . . .

SIZES OF THE CONTINENTS

Sometimes Europe and Asia are counted as part of a single large continent, Eurasia. Australasia (Australia, New Zealand, Papua New Guinea and other Pacific islands) is smaller in area than Europe.

	sq km	sq mi
Asia	43,610,000	16,840,000
Africa	30,335,000	11,710,000
North and Central America	25,350,000	9,790,000
South America	17,610,000	6,800,000
Antarctica	13,340,000	5,150,000
Europe	10,500,000	4,050,000

Acknowledgements

Heather Angel, Aquila Photographics, Ardea, Associated Press, Graham Bingham, British Museum, British Arctic Survey, Camera Press, Bruce Coleman, Colour Library International, Colourviews Library, Daily Telegraph Colour Library, Mary Evans Picture Library, Feux Rodriguez de la Fuente, Archivio IGDA, ICP, Jacana, Roger Kohn, Leeds Infirmary, Marka, Musée des Techniques, National Maritime Museum, National Portrait Gallery, Photo Source, Picturepoint, K G Preston Mapham, Karen Scott, Spectrum Colour Library.